PRAISE FOR HOW HARD CAN IT BE

"It's not easy to openly share your own failures. Arnaud writes about his journey in such a compelling and honest way that he gives the rest of us permission to do the same."

— **SUNEEL GUPTA, CO-FOUNDER OF RISE AND AUTHOR OF *BACKABLE***

"*How Hard Can It Be* is a real page-turner of a book—a fascinating and highly readable account of all the trials, tribulations, pivots, reversals and assorted near-disasters involved in launching and monetizing a startup business."

— **DON PEPPERS, AUTHOR, SPEAKER AND CX AUTHORITY**

"In this intimately personal account, the author takes us on a journey through the birth and death of a dream. *How Hard Can It Be* is more than a story of a business venture; it reveals the complexities and messiness of what it means to be human in an enchantingly honest way. A must-read for budding entrepreneurs as well as people interested in life, love and learning."

— **SUSAN GOLDSWORTHY OLY, PROFESSOR OF LEADERSHIP AND ORGANIZATIONAL CHANGE AT IMD BUSINESS SCHOOL AND AUTHOR OF THREE AWARD-WINNING BOOKS INCLUDING *CARE TO DARE* AND *CHOOSING CHANGE***

"Arnaud captures the emotional roller-coaster of a startup entrepreneurial journey. No amount of 'how-to' guides can truly prepare founders for the turbulence, anguish, elation and surprise—all in a single day. His personal story brings this to life vividly."

— **AZRAN OSMA RANI, FORMER CEO OF AIRASIAX AND SERIAL ENTREPRENEUR**

"Failure. It's something that we all try to avoid, yet experience nonetheless. How Hard Can It Be is a book that finally brings failure to the surface, to dissect, examine, and learn from. If you want some clues on how to best metabolize failure - read this book!"

— **RICH DIVINEY, RETIRED NAVY SEAL COMMANDER AND AUTHOR OF BESTSELLING BOOK THE ATTRIBUTES**

"Learning is the only way to turn failure into success. Arnaud's story of the ups and downs of his entrepreneurial adventure is full of lessons. He does not glorify failure but treats it with the respect a good teacher deserves. His narrative style is a refreshing new approach to the tired management genre."

— **DR. SAMUEL WEST, FOUNDER OF THE MUSEUM OF FAILURE**

"It takes a lot of courage to share a failure story. Arnaud has done that in great detail. The price he has paid for that failure has turned into a book as a present for you filled with lessons to learn, and more importantly, mistakes to avoid."

— **RAJESH SETTY, CO-FOUNDER OF AUDVISOR.COM AND AUTHOR OF *SET-UP TO STARTUP***

"Arnaud has managed to convert his entrepreneurial story into a page-turner full of lessons. Get ready to learn about the real day-to-day of startup life."

— **PEPE VILLATORO, CO-FOUNDER OF FUCKUP INC.**

"In professing his failure in a most personal manner, Arnaud asks us to consider why, when the vast majority of startups fail, so few accounts of failing are available. With so much to learn from failing, maybe, just maybe, the time for more failures to step forward and share their experience is upon us."

— **MATTHEW HIGGINS, SENIOR LECTURER AT THE OPEN UNIVERSITY BUSINESS SCHOOL**

HOW HARD CAN IT BE

Startup Lessons From Trying
(And Failing) To Take Down Facebook

ARNAUD HENNEVILLE-WEDHOLM

LEGACY launch pad PUBLISHING

ISBN: 978-1-951407-91-9 (paperback)

ISBN: 978-1-951407-90-2 (ebook)

DISCLAIMER

Arnaud Henneville-Wedholm asserts the moral right to be identified as the author of this book. The author does not make any guarantee or other promise as to any results that may be obtained from using the content of this book. Content contained or made available through this book is not intended to and does not constitute business advice. The author is providing this book and its contents on an "as is" basis. The use of the information in this book is at the reader's own risk.

While the story in this book is true, it is to be noted that the names of some of the characters have been changed for privacy purposes. Similarly, some names of organizations, corporations and institutions may have been changed. Finally, some events may have been altered to offer a deeper reading experience and an easier transfer of knowledge.

In loving memory of Lovisa

If you enjoy *How Hard Can It Be*, please leave your honest review on Amazon or drop some kind words on your preferred social media platform. It may feel unimportant, but doing so **helps more people** discover the book and its lessons. The more visibility *How Hard Can It Be* gets, the higher the probability more startups survive the tough entrepreneurial journey—and that we all win in the end.

CONTENTS

A WORD FROM ME

The bulk of startup stories the world is familiar with tend to end with millions of dollars in the bank for the founders after an oversubscribed IPO. Yet despite the ever-growing torrent of inspirational stories and the myriad of how-to books flooding the internet and filling the business and self-help sections of bookstores, startups are faced with a harsh truth: irrespective of a startup team's ideas, talent, funding, hard work, ambition or business milestones achieved along the way, nine out of 10 startups still fail, according to a 2019 report by The Startup Genome.

It is this simple fact and the question of what could be done to remedy this obvious problem that has haunted me ever since my own startup slid into its own downward trajectory almost a year before collapsing for good.

Clearly, it is not that the spectacular accomplishments of Jeff Bezos, Oprah Winfrey or Elon Musk, to name a few, are not relevant or helpful; undoubtedly, their fascinating stories deserve to be told. Even so, can the glorified accounts of stellar growth startups and their founders' incredible stories help the rest of us succeed in our own comparatively mundane entrepreneurial endeavors? When was the last time you applied a lesson you learned from, say, the biography of Steve Jobs? Or a lesson from the life story of any other A-league entrepreneurial player for that matter?

These stories, I believe, are to everyday entrepreneurs what a late-stage Picasso painting is to a fledgling painter or a Beethoven symphony is to a budding musician. Despite what we may want to believe, masterpieces are great for inspiration but are ultimately not great for learning. If you don't believe me, take it from Bill Gates: "It's fine to celebrate success, but it is more important to heed the lessons of failure."

My hope in writing *How Hard Can It Be* is to deliver a no-nonsense chronicle of my own startup journey that can somehow limit or lessen the amount of startup failure precisely by embracing Gates's prescription. In these pages, I offer my complete understanding of what took me and the rest of the challengesЯus.com team years to uncover in the hopes that younger entrepreneurs can learn some essential concepts to survive the wild rollercoaster ride of startup life (and avoid some of our pitfalls).

This book will reveal the process our team came up with to generate ideas worth pursuing. It will explain the importance of a balanced founding team and how to crack open that hard nut of scalability. Lessons on product-market fit, pivoting and positioning are all here as well—along with my team's sometimes hilarious stories of raising capital and acquiring customers. As the story unfolds, my hope is that readers will come to realize that "hustling" alone is rarely the answer.

I could not have written a book about the startup journey without talking about motivation, mindset and resilience—and while understanding these topics will undeniably set the most ambitious individuals up for success, one's ultimate future competitive advantage will stem from the understanding that the line between success and failure is *really* thin.

They say to write the book, you need to read. In that spirit, I submit that I wish something like *How Hard Can It Be* had existed when I started out. All the same, I also recognize that I do not have all the answers to guarantee anyone the success they deserve. No single individual can make that promise. Collectively, however, we might reach a better answer.

My ultimate vision is for this book to serve as a platform, a sort of "call for humility" for entrepreneurs to come together and help one another: not with examples of their successes but by revealing their failures. If entrepreneurs can honestly share their insights and hard-earned wisdom, we as a community might also dramatically increase the odds that the newest in our ranks become more than roadkill on the highway of innovation and progress. And if my own transformation into roadkill can be the catalyst that sparks that change, I'll be more than happy to have played that part.

With gratitude and no further ado,

Arnaud Henneville-Wedholm

Learn from the mistakes of others. You can't live long enough to make them all yourself.

—Eleanor Roosevelt

1. GOING LIVE

Stockholm, February 2, 2013

My laptop is open. I am on a Skype video call with Magnus, my co-founder, and our two programmers, Frida and David. This time is not a working session; it's the real thing. We are waiting to go live.

The plan once that happens is to celebrate, though logistically, I am wondering how we will accomplish that given that we are all in different locations. I am in Stockholm, where I live, while the rest of the team is in Uppsala, 60 or so kilometers away. We were supposed to be together for this important milestone, but for numerous reasons we are not.

On my side, I have some bubbly chilling in the fridge. My wife, Claire, knows I don't like champagne—and as a matter of fact, neither does she. But when I told her we would definitely go live today, she insisted on stopping by the liquor store on the way home from work.

I check the time and see that we've been postponed. Again. We were supposed to have hit "enter" at 5 pm, but here we are at 6:45 pm, still on standby. David assures the team that there is just a tiny fix—it's "the blockers" (whatever they are).

"How quickly?" I ask impatiently. "Five? Ten minutes?"

He laughs. The call reception is so bad that his voice sounds metallic, nearly Marvel-comic villainous.

"No," he finally replies in his serene and nerdy voice. "More in the neighborhood of two hours."

"Are you kidding me?" I respond with a slack jaw. I am now visibly deflating, although David is still calm and determined to move forward. As I've learned over the past year, he and I occupy opposite ends of the emotional spectrum. I can hardly, if ever, emotionally contain myself; he, on the other hand, is the kind of person the word "unflappable" was invented for. Even on launch day, minutes (well, now hours) away from finding out whether we have built a unicorn—a horse with a horn strapped to its head, or just an ass—he is entirely self-possessed.

I keep playing through the same two opposite-extreme scenarios in my head: either we go viral immediately and Mark Zuckerberg is on the phone at the end of the week to acquire us, or we nosedive from day one and watch helplessly as the last 14 months of hard work and sleepless nights get flushed down the toilet. Stepping out of my own emotional volatility for a moment, I remember that David's blank affect is part of what makes him so good at what he does. It's also why we brought him onboard in the first place.

Internally, I call David "Napster"—not because he is a multi-millionaire like the oh-so-famous founder of the 1999 peer-to-peer file sharing platform, Sean Parker, but because he is a real-deal computer hacker through and through. Although he doesn't say it (he rarely says much at all), I know he privately wears the nickname like a medal.

Sure enough, Napster's self-imposed deadline of two hours blows by, and the next time I check the time, it's 9:38 pm. As far as I can conclude from David's increasingly heavy breathing, we are nowhere near hitting the fabled "enter" button.

I'm using the downtime to plaster Twitter, Facebook and every other available social media channel with inspirational quotes from world-class achievers. We don't really have a scripted marketing plan ahead of our big launch—at least, not the kind one would typically expect from an established business.

That said, we collected tons of material over the previous months to scatter on social media throughout our launch—from one-liners to clever memes. We even got footage from numerous elite performers spanning everything from art to sports. My favorite of these is a video of a few Red Bull– sponsored base-jumpers in wingsuits hurtling into the "Crack" gorge in Switzerland like fearless flying opossums. It all followed a decision we made early on (you could call it a strategy, if you were feeling generous) to bring variety and eclecticism to our marketing. We were certainly going to be eclectic; what happened beyond that was anyone's guess.

I was drawing lessons from a book I'd read called *Differentiate or Die: Survival in Our Era of Killer Competition* by Jack Trout. To embrace the idea of survival through difference, we chose orange as a central color to punch through the various blues of every cluttered social media feed, along with contrasting black, gray and white (in no specific order).

All our marketing images show people in action, but the additional layer is that all the heroes are everyday people doing extraordinary and mundane things alike. Our idea was simple: anyone looking at the images should be able to relate to at least one of the situations they are presented with. Some of the images are really spectacular, such as one gent standing on a mountain top owning the moment, or a young girl in her late teens doing a Crooked Grind on a graffiti-covered half-pipe. Others are deliberately more accessible, such as a roller coaster plunging into a void. It looks thrilling, but it's still easy to imagine yourself sitting in the cart.

Our target market is the whole world—everyone in search of some inspiration to go beyond their comfort zone, at least. Our hope is that our market amounts to a lot of people, which is why we've loaded our marketing guns despite being on a shoestring budget. We have some 30 days of media ammunition at a clip of one new quote, image and video a day. It is a lot of content, but will it be enough? No one has a clue.

Earlier in the day, I reminded friends and everyone who volunteered to be part of our beta-testing three months earlier that our launch was imminent. It was maybe 400-odd people ready to challenge each other and jumpstart our movement. But the delays were threatening everything.

"Guys, we can't wait any longer, can we?" I asked. "We are going to look like fools."

I am hoping Frida will say something—something that will make Napster find a shortcut. Still, she has left her video frame—she's on mute, and the only thing I can see in her square is a white Ikea cupboard I've become accustomed to. After being surprisingly silent for hours, Magnus finally speaks up.

"The team has been waiting an extra four months," he says, "so a few more hours won't make much of a difference, don't you think?" Magnus is a man of few words, but his words are spot-on as usual.

There is a slightly uncomfortable silence as we listen to David pressing computer keys. When I least expect it, the word "done" echoes through the room, and David faces the camera with a big smile on his face (rather unusual, I must confess).

"Just say the word, and we go live!" he yells euphorically.

Almost in unison, Magnus and I scream back at him, "Go, go, go!" as if we were both part of the same Navy SEALs training. With the damn "blockers" removed, we emerge from stealth mode and the "coming soon" page we're all so tired of is magically replaced by a sign-up page. It's 10:12 pm and challengesЯus.com (a terrible name, but the only thing we could all agree on after months of debate) is officially open for business!

Finally, the four of us can celebrate.

We roar and laugh frantically. It's a surreal moment that's part exhilaration, part relief and part anxiety. What happens to us next is unpredictable and up to the invisible hand[1] of the market; still, there's no denying that this is a huge milestone.

The team agrees to hang up and reconvene in one hour as I collapse on the couch, and Claire brings over two flutes of champagne.

"Enjoy this moment," she says as we clink glasses. "You deserve it." In all humility, she's probably right—everything we accomplished over the last 14 months was done while working a regular 9-to-5 job (8-to-7, technically).

I try to cherish the moment. Our kids, six-year-old Chléa and three-year-old Melvin, are asleep. Tomorrow is Saturday, which means no office to go to—*and now that we've launched, no office to ever go back to*, I secretly hope.

I hear my phone vibrating, and Claire hears it, too.

"Go for it," she says resignedly. It's Magnus.

"Dude, where have you been? I've been calling you non-stop for a good five minutes." He is totally out of breath and sounds almost crazy.

"Me? I'm good, man," I say. "Just chilling over here."

"Chilling?!" He shouts incredulously. "*Är du* out of your mind?" I know he is excited because he's speaking quickly and inserting Swedish words into the middle of his otherwise perfect English.

"*Har du sett* the numbers? You have your computer nearby?" he says. Now I'm freaking out. Are people signing up, or is it dead quiet? Did the server go down? Are we already out of business? I grab my laptop and go straight to the analytics page.

"Holy shit," I hear myself say as I call Claire over. "Can this really be?" The real-time feed of sign-ups reads 2,735, and we have been open for under an hour. I grab my calculator to verify, and I cannot believe it—we are signing up some 45 users a minute!

"At this pace and considering the viral mechanics we've built in," Magnus says, "we will have tens of thousands of users by the time we wake up tomorrow."

"That can't be true!" I respond in disbelief. I am usually more optimistic than Magnus, but he's showing me up here.

I am not sure to which of the last statements Claire agrees, but she nods.

"Babe, it's working!" she says (a pet name she uses for me only when I'm in her good graces). Still, I don't know if it *is* working yet. I'm second-guessing everything. I had suppressed my hopes so as not to be disappointed if everything blew up; now, I realize I'm entirely unprepared for something good to happen.

To investigate further, I turn to Twitter. I am gobsmacked. In 10 minutes, we have been retweeted hundreds of times. We don't know much about Twitter—in fact, we know nothing about it—but if something gets hundreds of RTs in minutes, surely that's enough to go viral, no? I scroll through what people are tweeting about us:

The real social media for people who want to DO things.
This is a revolution!
Bring on the challenges.

Journalists from Silicon Valley and Israel, two of the most respected tech centers in the world, are already positioning us as "the next big thing." I relish the lede in one of the pieces: "A hot new social media platform that mixes the virtual and real world just launched out of Stockholm."

Young girls from places as far-flung as Uzbekistan are tagging their friends and challenging them directly. On Facebook, we see the same thing. Our likes and shares are going through the roof, and comments are pouring in.[2]

As I gaze at our site, new flags are popping up from across the globe: Mongolia, Belgium, Peru, Croatia, Australia, the Philippines. Magnus and I are dumbstruck. *We really did it.*

But I'm getting ahead of myself. To start at the true beginning of my entrepreneurial journey, I must return to my early 20s in France, where I met the love of my life: Claire.

2. MEETING CLAIRE

Claire and I met in June 1996 in Marne-la-Vallée, France. Though most people have never heard of it, in 1996 it became host to the newly opened Disneyland Paris, the new European home of Mickey Mouse.

We were in our early 20s and had both secured summer jobs at the park, among very few young people selected to deliver (and live) such a unique experience. Claire had come from Uppsala, Sweden, the fourth largest city in the country, to France to practice her French. Fortunately, she was already rather proficient upon arriving—which was good because back then, my English sucked, and my Swedish was non-existent.

While being from two different countries in a relationship certainly felt exotic during the summer, we soon found out how difficult it would be to maintain. Before I knew it, August was behind us, and Claire had returned home. The magic of Disney had vanished, both literally and figuratively.

We agreed to keep in touch, and to this day, I still remember pushing through the heavy glass doors of the futuristic-looking internet café at the corner of Rue Oberkampf and Boulevard Ménilmontant, in the 11th arrondissement of Paris, to open our very first Hotmail accounts together so we could stay in touch.

We started dating long distance and blew a small fortune on travel and prepaid cards for international calls, though we only got to see one another twice a year since money was tight. Things began to change when I got a call from a recruiter offering me the chance to manage my own gym in Lille, located in northern France.

As the recruiter explained, it would not really be my own per se, but I would oversee starting it from scratch—post-construction but pre-opening. By that point, I had been working in sales within the French fitness industry for several years and was consistently exceeding my quotas; what I had not realized was that my name had started circulating.

I thought of a famous quote from Richard Branson, one of many entrepreneurs I idolized: "If someone offers you an amazing opportunity and you are not sure you can do it, say yes. Then learn how to do it later." It was advice I decided I would follow to the letter, and I accepted the job.

The gym was quite impressive. In all, it was 3,000 square meters and had a pool, several studios for group activities like spinning and even a full thalassotherapy[1] corner. There were racks of shiny iron weights, rows of ellipticals, joggers and all sorts of other pain-inflicting machines I had learned to love. As we matured, I brought in 15 employees, from sales reps to fitness instructors, and we hit more than one million euros in revenue within the first six months of operations, which felt huge at the time.

After our summer together had ended, Claire continued her first-degree university studies in Uppsala and was on her way to a master's degree at the University of Kent at Canterbury in the UK. She'd done so to ensure we could see each other as often as possible since Canterbury was not too far from Lille. Thanks to the Eurostar connection between Lille and St. Pancras in London, we were suddenly able to see each other every third week, which was a huge improvement compared to just twice a year. With Claire graduating soon, it seemed like we would soon get the chance to be together in the same place, which I couldn't wait for.

Unfortunately, my dreams were dashed when Claire graduated in the summer of 2001 and was offered an internship at the United Nations in Bangkok. When she told me, I almost couldn't believe what I was hearing. Bangkok, *Thailand*? That was a hemisphere—not a channel—away. Even so, the opportunity was undeniably important for her, and I didn't want to hold her back.

The period that followed was the toughest we had been through yet. There was a substantial time difference between us, and we could seldom talk. We still emailed as we promised we would, but not being able to plan when we could see each other again was particularly depressing.[2]

Six months later, Claire secured a proper contract with the UN in Bangkok, and I realized there was no more time to waste. I had to act quickly if we were to continue our international love story.

Without too much hesitation, I quit my job at the gym to follow my heart and landed in Bangkok on August 5, 2002, with only a bag on my back.

After settling into our apartment, all I could focus on was how amazing it was just to be together. Overnight, I had gone from easily clocking 80 hours a week managing the gym and subsisting off of canned ravioli in my apartment to becoming a *farang*[3] in flip-flops meandering in the streets of Bangkok. I often took long walks through the city to breathe in the unusual smells of overripe durian fruit and the celebrated local *khao phat.*[4] When I wasn't savoring the flavors of this new world, I was bumming around by the pool at our apartment.

From my near-daily commutes on noisy express boats across the Chao Phraya (the "River of Kings" that runs through Bangkok) to the many lunches Claire and I shared at the local restaurant next to her office, one thing was clear: life was awesome. All the same, my Bangkok honeymoon was coming close to its end.

I started losing my Zen when I realized it would be virtually impossible for me to develop my career in the

"Land of Smiles."[5] Jobs were listed as for nationals only, and there was no possibility of getting a work permit. Because I had no intention of growing dreadlocks and smoking weed all day like Leo DiCaprio in *The Beach*, I was left with only one option: go back to school. In thinking of where I would go to school, I tried to think strategically.

After the dot-com bust and post-9/11, growth in the world economy had largely plateaued—but the same did not seem as true for the so-called Asian Tigers.[6] Though Asia had just gone through its now-infamous 1997 financial meltdown, the region was on an impressive rebound and still seemed to have long-term growth potential. China and India were averaging an annual GDP growth of 10.5 and 7.3 percent, respectively, and Hong Kong, Indonesia, Korea, Malaysia, the Philippines, Singapore and Thailand were all doing reasonably well, too.

Naturally, with so much economic transformation came optimism; foreign direct investments were pouring in, and with them, businesses were popping up everywhere at an ever-increasing rate. Part of that globalization included Britain exporting something more valuable than their pubs, car parts and broadcasting equipment to Asia: their education.

Higher education in Asia was like a new gold rush. British universities, it seemed, were more efficient at opening local versions of themselves outside of Britain than

McDonald's was at copy-pasting its franchises around the world.

For established universities, Asian satellite campuses meant lower labor costs for salaries and no centuries-old buildings to maintain. It also meant students could enroll in a world-class MBA in Bangkok for around half of what it would cost in the UK. The more I thought about it, long-term relocation to Bangkok started to seem like a very good move.

With Claire now holding a master's and working for the UN, I was feeling somewhat academically behind. It was a familiar sensation; after all, I had generally followed the path of least resistance throughout my childhood, much to my mother's chagrin.

"School will set you free!" she used to tell me. My mother taught mathematics at the local college in Montigny-lès-Cormeilles, a commune in the northwestern suburbs of Paris, and I remember ironically telling her that I doubted what she was saying was true. After all, she was still stuck in a school herself and was now always complaining that her students "were not getting it." To me, it sounded like a life sentence.

Over the years, my school report cards provided consistent feedback: "Arnaud is passing, but he could do so much better. What a waste of potential." After my baccalaureate (which I obtained without studying much at all), I announced that I would study tourism in college, which my parents met with quizzical looks.

"If I am to be truly free," I explained to them, "it's the only subject that has any potential at all." My father laughed. My mother just rolled her eyes.

I thought tourism would suit me at the time because it was not especially intellectually challenging. Although that might've been true to some extent, I must admit that the required courses in Egyptian, Roman and Greek history—the wars, the art, the architecture, the geography and the rest of it—were far more interesting than I had expected. Languages were also an important part of the curriculum (a blessing, since it was one of few things at which I was academically gifted), and I simultaneously learned English, Spanish and Italian. Regardless, having a good time in the sun on the beach was my real motivation—not mastering Mesopotamian history or learning Amadeus.[7]

After earning a bachelor of arts in 1996 with a specialization in the "commercialization of tourism products," I realized I was only fit to sell dreams to others, in practical terms, but I had no intention of being stuck behind a desk flashing fake smiles for the rest of my life. It was why I quit the tourism business before even getting started and, after a year as a headhunter and 10 months of military service, I joined the fitness industry in 1999.

Fitness had been much more enjoyable, but it wasn't doing anything for my intellectual insecurities.

Since I was living in one of the most eccentric cities in

the world, I took aim at getting a new education—a real one this time. I considered two universities with programs in Bangkok, namely the British University of Leicester and the Edinburgh Business School. Neither one was cheap, but the latter turned out to be exorbitantly priced, so I chose the former.

On the first day, reality quickly caught up with me. While my English had sounded reasonable in a French context, it did not translate to an international one. For the first several hours of my class, I sat listening to my Australian management professor who had somehow made it to Bangkok from the outback. The only thing I understood was when he said, "G'day!"

Claire and I lived on the 17th floor of a high-rise building in central Bangkok. Our flat was huge—at least compared to the apartment I'd been renting in the north of France—but the internet was terrible. On a bad day, it could take up to 20–30 minutes to have the 56kbps dial-up modem connect to the World Wide Web. The slow speeds drove me nuts, largely because most of my assignments had to be submitted by email. Fortunately, all my classes were conducted in an actual classroom.

After looking back on my academic achievements, I intended to correct my trajectory by getting up to speed with English (which cost me about half a year of daily sweat). In school, I found once again that sitting on my ass day after day to flip through theories was less than stimulating. I enjoyed the ideas for what they were, but I saw myself as a man of action. I needed to take a leap of faith and put the theory of entrepreneurship into practice.

* * *

Anyone who has ever been to Bangkok knows that it is essentially one giant neon billboard. Walking around, you are surrounded by countless commercials and TV screens blinking and blaring at you, day and night.[8]

Ads were pervasive, from giant stickers glued to the pavement to conversational barcodes in local supermarkets to screens attached to every standing pole of every train's carriage. They were in places you would least expect them, like on top of toilet seat covers at bars and restaurants.

The problem with old-school ads was that their effectiveness was very tricky to measure—and without any real evidence to go on, the best strategy was generally "the more, the merrier." One thing that was certain, however, was that after seeing the same advert over and over, people stopped acknowledging it. *Instead of plastering billboards everywhere with nonsensical ads that people ignore,* I wondered, *what if you were to get the message directly in front of the right people instead?*

As I mulled it over in the back of a cab one day, on the way to finance class and stuck in Bangkok traffic, I realized there was still one place that wasn't being exploited for ad placement: private cars. By the following week, I had registered a company and was thinking my idea through, day and night.

After a few months of brainstorming, I outlined two types of services: one for consumers and one for enterprises. The company I imagined would wrap private cars with the kind of perforated see-through films sometimes found on buildings, but this would be combined with geomarketing. Private customers would register their cars with the company to carry advertising; in return, the company would pay them in hard cash, which I estimated would largely cover monthly petrol and insurance expenses for the best carriers. After all, who didn't want to drive for free?

Based on various criteria, advertisers would then be able to select the kinds of cars they wanted to carry their brand and to literally "drive their message home." It was brilliant!

To get the company off the ground, I joined forces with another *farang* named Günter, who was a resident of Thailand, a fellow MBA student and a fluent speaker of Thai. When we had enough cars registered, we knocked on big advertisers' doors. It worked like a charm; consumer products, airlines and shopping malls were among the first to jump onboard.

Our second offering was an ad-wrapped fleet of cars that we called "caradvans." Our caradvans would drive wherever marketers wanted them to, although we also had predetermined routes at different price points from which they could choose. Each route had specific stops where hostesses in flashy outfits would jump out to distribute leaflets, among other enthusiasm-generating stunts. It was "below-the-line" marketing,[9] though I preferred the term "guerrilla marketing" because that was who we really were: guerrillas in the jungle of Thai advertising.

Neither Günter nor I had any idea of what to do or how to do it, but we knew we needed a supplier willing to take risks with us. Even so, risk-takers were not exactly queuing up. Our printer would be more like a partner, and although we did not have anything to offer upfront, we were trying to pitch the fanciful idea that by working with us, they could add to their topline and perhaps find riches through diversification. Günter and I had visited quite a few printshops before agreeing to work with one (more accurately, before finding one that agreed to work with us).

Our contractor was a small printshop lost somewhere in an alley branching off a major road, and they had never printed anything to stick on a car before. Because they had no extra space where cars could be brought in for the "wrapping" process, we concluded that we would have to wrap at a different location of the car owner's choosing. It would be a better deal for the clients anyway.

Our first client was a local chain of fast-food restaurants. We got the printshop contractors to meet us in a parking lot to do the job. And although we had agreed to meet at 6 am, the team showed up around 8 am (and to my surprise, at least 10 of them came along). The job was to wrap five company cars in 12 hours, and we had estimated the work at two hours per car.

As the team laid the stickers flat onto the bare concrete of the parking lot to figure out which part was to be stuck where, we came to realize there had been some "anomalies" in the printing process. Some stickers were poorly printed, others were wrongly perforated and some that were supposed to fit onto particular cars did not match at all. In the worst cases, some parts of the actual ad were missing. It was a bit of a disaster.

By noon, we still had not wrapped a single car. Clearly, we would not be able to fulfill the delivery terms and have the cars ready by 6 pm that day. The client turned out to be kind enough to give us an extra day to complete the job, a deadline we barely made (even with the team working through the night).

Even if the first job had not been profitable, we had learned a lot. On the next one, we were determined to have a much better process: fewer errors and perhaps some profit. Slowly but surely we improved, one client at a time, until we became quite good at wrapping cars—we even decorated cars carrying people to the Bangkok Zoo like zebras, and other cars advertising restaurants like rolling bowls of noodles.

For a while, business was booming—but after a mere two years in business, a large Thai advertising group entered our lane and crushed us overnight. As our revenue dried up, we suddenly realized there was nothing we could do but fold. We did not own anything. We had no production; we were simply middlemen who were outsourcing the whole process of design, printing, car wrapping and film removal to a local printing shop. The Thai company, on the other hand, had it all: the capabilities, established customers they could pitch to, the deep pockets, the connections to the right people in positions of power and so on.

Looking back, Günter and I had largely pioneered a niche. We were certainly the first ones doing mobile advertising on private cars in Thailand, even if it had ended in defeat. It set an important precedent that it was possible to start your own business, give it your all and have it become a success. It was my first real entrepreneurial experience, and it had planted a seed. The rest of my education was rather uneventful by comparison.

I graduated in June 2006 but did not go to the hat-throwing ceremony. It felt a bit excessive to pay all that money to fly out to Leicester for a mortarboard muck-in, so I stayed in Bangkok. Instead, Claire and I went for a luxurious all-inclusive weekend on the beach. It felt appropriate because we were also celebrating our return to Sweden, which was less than two months away. Our time under the palm trees, it seemed, was finally coming to an end.

3. COLE+MOORE, MAGNUS AND I

At 6:45 am on July 15, our Thai Airways flight number 731 from Bangkok hit the ground in Stockholm. After a 10-hour flight, we were finally on Swedish soil. It was an exciting new beginning since it would be my first time living and working in Sweden. Until then, I had only visited a few times.

It was also a new adventure for Claire and me as a couple, though soon we would be more than a couple. Claire was pregnant, and our first child was due in November, which for some reason I assumed would be a boy.[1] When I shared the big news, a close Italian friend of mine named Alessandro told me an interesting saying from his village in Calabria, Italy's southernmost region. In not-so-grammatically correct English, he said, "All mens want boys, but good man have girls."

After his comment, I was crossing my fingers for a girl

—though ultimately, I would be happy in either case. Claire and I wanted it to be a surprise.

Upon arriving in Stockholm, the first job I landed was at the marketing department of a leading global consumer product company. It was a good start but not especially intellectually challenging (which was a little disappointing after studying like a maniac for over three years). When the opportunity to join a consulting firm presented itself in August 2007, I quickly accepted.

The company was Cole+Moore, a high-end boutique management consulting firm that was a second-tier company rather than an industry leader like McKinsey, BCG or Bain. While strategy reports and other sorts of high-level analysis are typically done by the "big three," Cole+Moore's focus was on strategy implementation.[2]

Cole+Moore's focus was on behavioral implementation,[3]and we often joked that unlike McKinsey, BCG and Bain, we were the ones who did the real work.

My job as "hunter-in-chief" was to go after large companies that Cole+Moore wanted to have on its brag-sheet. I oversaw new accounts, mapped them out, tried to understand how they operated and then identified and tracked down their executives. After that, I would put some of them "on a drip," which meant sending them emails with information Cole+Moore assumed they would find valuable at a regular cadence.[4]

To my baby-boomer bosses, the entire process was nonsense—a waste of time compared to getting clients to "sign on the dotted line," as was the axiom at Cole+Moore. My colleagues and I regularly laughed at that one-liner during *fika*,[5] an oasis in our otherwise secular Swedish work lives.

As I knew, the world was changing. Thanks to the internet, business was no longer about calling people to sell them stuff. Information access was no longer asymmetrical, and as a result, buyers now often knew which products and services they needed in advance *as well as* the details about what vendors were trying to sell to them; this was true at the business-to-business (B2B) level, as well. In light of those changes, any informational asymmetry in those exchanges that still existed was now mostly a result of laziness.

At major companies, switchboards were gradually vanishing, and it was becoming increasingly difficult to get in touch with specific individuals. Buyers were taking charge of the process, slowly but surely. For us salespeople, it meant that we had to become experts at reeling in the right buyers at the right time. It also meant that the relationship between sellers and buyers would now be built solely on trust and soft skills rather than establishing authority by rattling off a list of tech specs.

* * *

Magnus was a senior analyst at Cole+Moore. His job was to gather information about various companies—typically from their annual reports, industry-wide data and trends but also from interview summaries our own consultants had conducted with management executives of various top companies. After that, he put reports together based on data culled from those many disparate sources.

He belonged to the 10-person Development Team and had ”senior” in his title because he had already been working at Cole+Moore for around two years, having joined right after obtaining a double master's in economics and computer science from the University of Uppsala.

I was barely two days into my role when Magnus and I ran into each other—quite literally—in the corridor. He was fully immersed in his computer with his head down when he blazed straight into me. His large rectangular glasses went one meter above his head as my double espresso arced through the air, finishing its course on an immaculate white wall reserved for customer testimonials. We both apologized, introduced ourselves and moved on as if nothing had happened.

There were four departments at Cole+Moore. Other than the development team, there was the delivery team, composed of some 40 consultants who were out of the office most of the time (generally either in workshops with clients or on planes to workshops with clients). There was also the sales team, which was primarily the domain of senior management and was true of the entire consulting industry (the "alpha males," as they were often referred to).

Like most other consulting firms, the sales strategy of

Cole+Moore was to "sell from within," namely to other parts of the same client-company or to other companies those clients were connected to. Among all these teams, I soon discovered that I was really the oddball; more accurately, I was a guinea pig.

At that time, the biggest recession of the modern era (which has since been topped by the one triggered by the Covid-19 pandemic) had just started in Europe. Companies around the world were becoming increasingly nervous, and budgets were drying up. Even so, Cole+Moore's CEO was on a mission to outmaneuver the competition and beat the downturn. Though times were tough, our firm had decided to go on the offensive.

Our building was in Östermalm, a posh neighborhood in central Stockholm, and was dignified despite being more than 100 years old and having no elevator. Magnus was on the second floor while I was on the third. On the third floor with me was a massive golden bell—and anyone who signed a deal got to ring it. It was always the same guys—the "closers"—who got the privilege of ringing that thing (though I was thrilled to get the chance to ring it a few times myself over the years). It wasn't as big as a church bell, but it still made a hell of a noise; every time a deal got inked, the entire building shook.

Eventually, Magnus and I built rapport and started getting lunches together. We mostly discussed strategy and change management, the details of our responsibilities at Cole+Moore. His favorite topics were finance and economics; mine were more along the lines of entrepreneurship and cognitive psychology.

Our conversations went beyond work, and I started to appreciate his character. Soon enough, we were profoundly enjoying one another's company. Eventually we were eating together almost every day and joked about starting a business together. I told him about my business venture in Thailand, and it was the first time I saw real emotions coming out of him.

"That's incredible!" He said. "I had no idea you were such a maverick." Though he was excited, the timing for another risky business venture was bad, to say the least. The debt crisis was in full swing, and after the collapse of Iceland's banking system, problems were spreading faster than any brainy economist could've forecasted. By 2009, Portugal, Italy, Ireland, Greece and Spain were all on the verge of collapse.

On top of those external factors, we both had young kids and mortgages to pay. Factoring into that equation was the fact that after my virtual graduation, I had worked hard to convince myself that my startup days were behind me. I had given it my best effort, but now it was time to follow a "normal" path—meaning, I suppose, to keep a salaried full-time job and have kids, a Volvo and a white picket fence. As a result, our passionate discussions remained just that: refreshing escapes from our daily routines.

After about a year of regular daydreaming, the discussion had extended beyond our 30-minute lunch breaks. We were now meeting after work and on weekends, and part of the reason was that Cole+Moore's strategy of aggressive expansion in a downturn was not really panning out. I was making some deals but not nearly enough to keep all the consultants busy. Eventually, the firm went on a cost-trimming spree, and the central functions of design, IT and finance all got their share of excess cuts—hell, even some of the consultants were being sacked.

As Magnus and I looked at the bigger picture, the economy seemed to have hit rock-bottom—and from there, it was only logical that it would eventually go up again. We were coming to the conclusion that if we could have a company that was startup ready by the time the economy picked up again, we would be in a very good place.

We accelerated our idea-generation sessions. We started thinking of ourselves as machines, programmed to meticulously search out the next big thing. Because Magnus and I were both generalists with diverse interests, we were open to any idea that made sense. Eventually, we came up with our own model (as befit our consulting background) to test ideas that could be worth pursuing. It was a very simple yes-or-no, three-step model, which made it incredibly effective.

First, we had the "non-negotiable" criterion: working on the venture for the foreseeable future (with a conservative exit horizon of 10 years) would have to be fun. Furthermore, the fun index had to be roughly equal for each of us, which made sense since we would both have to give the project our all to optimize our chances of success. It was the first major hurdle because Magnus and I didn't have the same definition of fun.

There were very few industries I found boring, but banking was undoubtedly one of them; Magnus, on the other hand, went bananas about cash flow statements and P/E ratios.[6] Similarly, topics like neurochemistry, which had me in a state of near ecstasy, left Magnus unmoved.

Once an idea made it through our fun criterion, it had to pass the test of scalability. Since our aim was to strike it big, scalability was an absolute must. It was a challenge we had witnessed firsthand at Cole+Moore.

When Cole+Moore had a client, we could try to grow their account from within, but that was really our only halfway sustainable strategy. The problem with that model is that when most of your entire organization is busy painstakingly servicing a few clients, there is virtually no room for growth. People, and the value you can extract from them, are essentially a limited resource; it was why Cole+Moore's idea of expanding aggressively during a contraction was kind of a smart move. The leadership team knew that the time would come when existing clients would start cutting their budgets, and we would need to have a backup plan of other revenue streams.

In simple terms, scalability implies that a company can increase its sales faster (ideally, way faster) than it must add resources to generate sales. In practice, scalability is a scale in itself. For example, a package delivery system is somewhat scalable because buying a few more delivery vehicles enables the company to deliver many more packages. Still, a telemedicine application on your phone is likely more scalable because it can increase its number of virtual consultations exponentially with only a marginal increase in the costs of IT infrastructure.

Finally, if an idea passed our first two stages, the final criterion was still the toughest. We had dubbed it our "internal-alignment compass." If you put your mind to it, getting through the first two stages wasn't that hard. But as a famous quote attributed to Thomas Edison says, "Vision without execution is just hallucination." To actually execute was the job of any entrepreneur, so an honest self-assessment of whether we had what it would take to execute our idea was the third bar to clear.

When an idea made it to the third stage, Magnus and I often had diverging views; still, it was a divergence generally shared by the business community at large. On the one hand, the likelihood that an idea would work increased significantly if the founders were intimately familiar with the industry they targeted. If someone had already spent five to 10 years studying a field, talking to people and building relationships, they had a clear tactical advantage over a freshman.

On the other hand, it was not uncommon to see total outsiders turn an entire industry on its head because they started with fresh eyes. They began their projects with a beginner's mind, without the limitations of any established belief system, and they often innovated and succeeded as a result.

To be pragmatic, Magnus and I concluded that even if we didn't have all the required competencies between the two of us to make an idea fly, if we could imagine "buying" those competencies—whether by hiring team members or using the right software—the idea would still pass the last step of our model.

After debating business ideas for nearly another year, Magnus and I still had not moved at all. Though we were in no rush, it also felt like we were falling victim to "analysis paralysis," a professional hazard we always warned executive clients against.

It was clear that we had to avoid killing the very passion that had animated us for going on two years. We had already killed some 40 rather good ideas—from an augmented reality geo-marketing app[7] to a cloud-based to-do list service for busy parents[8] to a crowdsourced service for dog-sitting.[9] Some of our ideas required too much capital to get started; others, we concluded, were too far out of our comfort zone.

Without any traction, we continued our jobs in the rat race—but it was getting harder and harder. Both of us were losing the fire we once had for helping companies execute their strategies. Instead, our hearts were swollen with the hope of exploring new possibilities.

There was one topic Magnus and I kept returning to in our discussions: Facebook. As professionals in the field of human behaviors, Facebook sparked our intellectual curiosity. Knowing the self-harm and the evils of the company itself, why did people still engage with it? Why was it so popular? What was driving most of our friends—and close to 200 million people a month around the world, at that time—to share the tedium of their daily activities with the rest of the world (and with a particular fetishization of the food on their plates)?

We understood self-aggrandizement and navel-gazing, and we could grasp the instant gratification of dopamine hits from the "likes" people were receiving—but still, Magnus and I were in awe of Facebook's growth trajectory. It seemed nothing like any textbook growth pattern we had seen. *People will eventually come to their senses and realize there are no long-term benefits of scrolling through endless digital nonsense*, we thought. *The market will correct, no question about it.*

That was when the epiphany hit us: what if we tried to displace Facebook? If Facebook was destined to fail (as we decided it was), why wouldn't we be the ones to bring about their end? It was a good start, but how would we do it?

Clearly, we were not the first ones to be hit by a similar spark of inspiration. Plenty of other companies had tried getting in the game with platforms offering smaller communities of tight friends, or within specifical more verticals like music, fitness or even sports fishing.[10] Most of those concepts had already died or were on the verge of dying. Even so, we felt that we had a shot at succeeding. *If something as inane as Facebook could become so popular*, we asked ourselves, *how hard could it be?*

The crude reality was that neither of us had ever even built a website, let alone written code, so we were facing one of the steepest learning curves possible. Regardless, we knew that successful entrepreneurs needed a deep sense of faith that no challenge was de facto impossible. After all, wasn't it Steve Jobs who notoriously said that only the people who are crazy enough to think they can change the world are the ones who do?

I loved the quote because it was so inspiring, so I tried not to think of all the nutcases who tried to change the world and died trying—surely they outnumbered the people Jobs was referring to. Regardless, we were now fired up by a mission that was way bigger than ourselves. We were all in.

Back in 2009 to 2010, Facebook was nothing like what it is today. There were no fake news controversies or data breaches. There was no AI, no face-recognition software, no Russian interferences and no sharing of Trump's tantrums. Most of it was about free online games, family picnics, pictures of kittens with party hats and ads—not as many as there are today but still plenty. It still seemed possible to carve out our own niche, so we started playing with ideas for our own social network. After roughly three months, we had zeroed in on a concept.

One piece of research we had become obsessed with

was that Facebook and other apparently "engaging" online platforms were doing the opposite of what they intended in the first place: they were making people feel more depressed and alone. Clinical depression was on the rise, online bullying was making daily headlines and suicides among young people were increasing. Our thesis was that these platforms were turning people into anti-social animals, sad beings who were losing touch with what it meant to be alive. Instead of bringing people together, Facebook and their ilk were tearing the world apart and making people feel bad about themselves by encouraging them to try and live up to impossible ideals.

It was now so easy to "like" things from the sofa that real-life civic engagement was at an all-time low. Magnus and I joked about people liking important causes like Médecins Sans Frontières[11] without actually participating in the organization in any other way. They probably didn't even care about refugees or struggling groups in the first place; all they wanted was elevated status among their friends and a higher score on Klout, a service that rated people according to their online social influence.

The world had gone too far. The result, we figured (we were on a roll now), was that people would not even meet in person anymore—why should they? They already knew far more about one other from the internet than they would have learned from going out together every day for months.

Our vision eventually became clear: we wanted to be the "anti-couch potato" network. We wanted people to meet new people in person and to have real conversations, to be more active and get out of their comfort zones. We wanted youngsters to go outside again and to do the things we did when we were young. We wanted people to live again.

Through our platform, people would capture all those wonderful moments and share them on our platform. And because we *knew* people needed a push to get started on their journeys of self-discovery, a "challenge" could be the triggering event. We would inspire them by offering personalized challenges—private or company-sponsored—that would match their profiles and interests.

We would amplify real-world experiences online, which we thought no one had ever done before. Our plan was grand, our goal was well-intentioned and our enthusiasm was unmistakable. We told our friends that Zuck should watch out because we were coming for him with the coolest social network the world had ever seen. We may have sounded crazy, but so what? We were about to change the world.

4. THE SUN ABOVE THE CLOUDS

It took the world about a decade to recover from the burst of the dot-com bubble. Everyone had lost their appetite for risk. Entrepreneurs went silent, as did the entire business ecosystem and in turn, the media: when the circus leaves town, there isn't much to report on. But around 2005, the tide began to turn. With the rebirth of Apple with Steve Jobs at its helm, new-age tech companies like Facebook, Airbnb, Twitter, Zynga and Snap were popping up every where and making headlines. The "cult of the entrepreneur" was back in full force, and the world's attitude had changed: this time, things would be different.

By that time, I had been working for five years since my virtual graduation, and Magnus had been at it even longer. I was done with the red tape and was dead tired of the backstabbing and back-scratching. I wanted to do my own thing and to "be my own boss" again as the adage goes. By talking about it so much, Magnus was becoming convinced that it was a good idea for him, too.

"We're going to make it big," I would tell him. "*Huge*. The only limitations here are the ones we're putting on ourselves." If we had tested our testosterone levels during this period, I have no doubt we would have been above the average range for our age.

We encouraged one another to dream big. I wanted to be on the Forbes 40 Under 40 list; Magnus was talking about having hundreds of employees and an IPO.[1] I had walked this lonely road before and had some sense of the hard work it would entail. I also knew that the business I'd launched in Thailand would never have gotten us any venture capital (VC) interest; it was way too local, way too labor-intensive and not bold enough to give them a big exit (which is what all VCs are after).

If we were going to take the route of courting VCs, we would have to think big—really big. It was the only way to secure truckloads of money and have someone else finance our dream. It was how we came up with the idea of pitching ourselves to Sting.

Sting is the biggest startup incubator in Northern Europe and the eighth most successful in the world in terms of output. It is also the undisputed leader in its category in Sweden and is annexed to Sweden's most prestigious university, Kungliga Tekniska Högskolan (KTH), also known as the Royal Institute of Technology (even the name is intimidating). Disruptive innovations are channeled from KTH to Sting, but that wasn't the only way to pitch them.

As of 2002, anyone could apply to Sting provided they had a good idea and met some basic criteria. Every year since, Sting has received more than 500 applications from ambitious entrepreneurs around the world wanting to put a dent in the universe. There is a "but," though, and it is a big one: less than 10 percent of all Sting applicants make it past the first interview. In the end, only around 30 startups per year make it to Sting, but those who do become clear candidates for tomorrow's success.

The first time I pitched what the next social network would look like was at Sting on a cold late Friday in October 2011. I went to the pitch meeting on my own because Magnus was busy. My slide deck was neat, my pitch well-rehearsed and my conviction steadfast. Ready as I was, I rushed out of the office around 5 pm to meet Sting's first gatekeeper. The business coach to whom I was expected to present, Peo, was an Information and Communication Technology (ICT) expert. Peo was probably in his mid-40s. As I learned just a few minutes later, he was not your average interviewer.

He had gone through the startup journey himself not once but three times, each time pocketing more cash for his exit. He had made so much money that he was financially independent. His life was now dedicated to enjoying his good fortune and taking in life's simple pleasures, but his passion for tech had not changed. As such, he was now "giving back" by passing on his wisdom to help others improve humanity through tech.

As I pushed the door open, I received one of the warmest smiles I had gotten in a long time in a work context—it was so unlike the dull and unanimated faces of the C-level executives in dark suits I typically engaged with while prospecting on behalf of Cole+Moore.

It was a good start, but Peo had not made his money by being a laidback teddy bear—he was sharp. For the next 100 minutes or so, he grilled me to the bone, deconstructing challengesЯus.com and making parallels and connections to other projects that seemed obscure. At times—quite often, actually—I had absolutely no idea what he was talking about. As I learned during our debate, Peo had been an officer in the army before becoming a successful entrepreneur. He spoke Swedish and English but also Russian, French, German, Italian and Spanish fluently. By the end, I was convinced he was nearly superhuman.

After I finished my pitch, there was a nerve-racking silence before the verdict came down: *no*.

"No?" I repeated, surprised by what I had just heard. In my mind, the meeting had gone well!

"No," he said once more. "It is a no, for now. I like the idea and I think you are on to something, but I see three problems."

"Three?" I repeated incredulously.

"Yes, three. The first one is that right now, this is just a very nice story. You have nothing besides a few slides.

Clearly, you've done the homework. You are passionate, I'll give you that—but that's about it." I could barely believe what I was hearing.

"The thing is," he continued, "we do not incubate ideas here. We incubate entrepreneurs who have a proof of concept. I'm sorry, Arnaud, but you clearly don't match the profile. Yet." His timid final "yet" felt like he was trying to make his harsh feedback land a bit better. I had barely recovered from his first jab when Peo threw his next punches.

"The second and third reasons," he said, "are that you have no team and no experience." His second blow left me nearly unconscious. *What the hell is he talking about?* I thought to myself. *Who does he think he is?*

"I'm sorry, Peo," I said, "but I think you've got that one wrong. The full team might be debatable, but as far as experience is concerned, we have plenty of it. Together, Magnus and I have more than 10 years[2] of high-end business experience working with top-tier companies and helping them align their people with their strategies. I also launched my own company years back."

I was hoping for a bit of rest and maybe even an apology. But an apology was not quite what Peo had in mind.

"That's all well, Arnaud," Peo replied, "but that won't help you much with this project. And by the way—some tech expertise might come in handy if you plan to build a tech company." With that, Peo had knocked me out. He was right. We had nothing. How did we even get in front of him? How could we, the strategy consultants, have gotten it so wrong?

"Just take your time," he said as he walked me out. "Do what you have to do, and as I said, you are more than welcome to apply again once you've come a bit further—once you have a complete team, an MVP[3] and some demonstrated traction. Maybe next year?"

I needed some fresh air—a lot of it. On a normal day, I would have taken the tube to T-Centralen, the main Stockholm train and subway station, before riding another train to my stop. But this was not a normal day, and I was up for a walk.

My 30-minute walk through Stockholm was enough time to replay the entire meeting in my head. What had I said? Had I just screwed it all up? How would I summarize what had happened to Magnus? What would he say?

I analyzed whether I had made any major mistakes, but nothing too awful came to mind. As for the actual idea, Peo, being an ICT guru himself, had clearly validated it—or was he just being polite? Did he really mean what he'd said at the end? How could we put together a team with the right competencies required to execute the vision? I didn't know any geeks, that was for sure—but maybe Magnus did? Also, what was the MVP he mentioned?[4] I had no clue.

The weekend was tough. After it was over, it was hard to concentrate at work and to stay positive in light of recent events. Magnus had been called in to tag along on a delivery trip to Japan for Nokia, a longtime client of Cole+Moore. He would be away for 10 days. Since I had met Peo on a Friday and Magnus had flown out on Sunday morning, I had not even had a chance to debrief him properly, let alone discuss the next possible steps of our project.

A few weeks passed and we were at a standstill. We were treating our acceptance into Sting as proof or validation that our idea was a good one. When it didn't come, we were unclear what to do next. We could not go to the bank and take out a loan to finance our dream; that was for sure. We were happy to put in all the time in the world plus most of our savings, but we were not about to refinance our mortgages for the project.

The idea had crossed my mind a few times, but it wasn't a decision I could bring myself to make; it would affect Claire as well Chléa, our daughter, and Melvin, our little boy and the latest addition to our family. Claire would never buy into it. Similarly, Magnus was by nature more conservative (or as he preferred to put it, "more rational"), so for him, it was a non-discussion. We simply could not risk our families' future for the dream of displacing Facebook. We were too smart for that.

* * *

A good three months passed since meeting with Peo. I had assured him I would keep him up to date with our progress, but I had not delivered on that promise. To be honest, there was not much to update him on; we had not moved at all. It did not help that both Magnus and I had gotten busy at our regular jobs and, since our energy levels were low, we had relegated our project to the backburner. Our night chats had stopped, as had our exciting weekend meetings.

At work one day, I came back from lunch to find an email from Peo. It was not for me directly, but I was copied:

Hi Henrik,

I want to connect you with Arnaud. He has an internet media product in the making called challengesЯus.com. This idea is a bit outside the Sting box but is interesting anyhow. They do not have a KTH background, but I think you two could have a very fruitful discussion. You are the right person to understand this area.

I hope you enjoy the discussion with Arnaud.

Cheers,
Peo

Although the email wasn't a complete gamechanger, it was at least a new hope. Clearly, Peo had seen something in our idea that was enough to warrant an introduction to whoever Henrik was. I quickly looked him up on LinkedIn and was impressed to see that the man had a few successes of his own. Now he was teaching entrepreneurship at KTH. I sent him an email, and we set up a meeting.

The following week, Magnus and I were off to KTH to meet Henrik. The meeting was scheduled for 6 pm, after his classes. Henrik had asked us to prepare something following the INSWIN model. Neither Magnus nor I had ever heard about INSWIN, but it sounded reasonable once we looked it up. It was an acronym for Intro, Need, Solution, Why, Investment and Next steps:

Intro – Who are you? What are you presenting? What is your expectation from the meeting?

Need – What needs are you addressing? How big is the need? Who has the need?

Solution – What is your proposed solution to the need, and what is unique about it?

Why – Why are you the ones to make this work? What makes you different, and why should anyone trust you?

Investment – What do you need from me to make this work, and what can we all get out of it?

Next steps – What action do you want to see happen after our meeting?

As we walked through the deserted corridors of KTH, we were nervous. It was February in Sweden, which meant it was rather dark outside—not to mention freezing. Eventually, we reached Henrik's office at the very end of a corridor down in the basement—not exactly where one would think to find the offices of one of the institution's top dogs (at least, if I had been the dean).

Henrik did not even have a single window in his tiny office, but for someone who did not seem to get much light, he looked surprisingly healthy. He certainly was not a typical professor. He didn't have a knitted jumper or a messy beard. He wore stylish glasses, a pair of jeans, a black Ralph Lauren polo and a brand-new pair of white Adidas sneakers. As someone who frequently hit the gym, I could tell from the circumference of his arms that we had at least one activity in common.

We quickly got into the meat of our talk and proceeded through the INSWIN model. Henrik seemed to like our idea right away, but like Peo, he was not naive. He told us about all the startups he had been involved in over the years and was very proud of the fact that he himself had never actually started anything.

"My strategy is just to unearth gems," he said with a sly smile. It was quite clever; I had to give him that.

Though Henrik was interesting, he also seemed like a bit of a show-off—and at times during the meeting, I caught Magnus rolling his eyes. Most refreshing was that Henrik did not beat around the bush. He had no problem talking about unpleasant topics and just said things as they came to his mind, without any filter.

"Guys, what is the one and only one thing all winning startups have in common?" He said once we were through with the INSWIN. "Any guesses?"

Magnus and I looked at each other.

"USP? Their Unique Selling Proposition?" Magnus offered.

"Wrong!" Henrik boomed. "The team!" He turned around and pointed to the empty wall behind him. "You guys see that shelf?"

There was an odd moment of stillness in the room. As Magnus and I said nothing in return, he asked once more with what sounded like slight irritation in his voice.

"You see that shelf behind me, right?" There was clearly no shelf on the wall, but Magnus and I both played along by nodding in unison.

"Well," he began with a grand flourish, "that shelf is full of crying babies abandoned by their parents." The guy was now officially freaking me out. Magnus was kicking me from under the desk as he looked at me with a stone face.

"You see my friends," Henrik continued, "this is how it goes. A lot of people have ideas. Sometimes they even have good ideas. After a while though, those same people who were once full of passion and drive and who wanted to change the world quit. They give up like the sissies they are. And what am I left with? A lot of whining orphans I don't know what to do with. And guess what? I am a terrible stepfather!"

Beyond his bizarre allegory was a plain, chilling question: What kind of entrepreneurs were we? Would we be like the many other overly ambitious youngsters he had dealt with?

There was not much we could hide from him. He had done his homework, and he had a good understanding of what we did for a living. He knew where we lived and zeroed in on our socioeconomic backgrounds with just a few educated guesses during our meeting. It was a bit spooky, in all honesty.

Besides being a professor, a fabulist and perhaps a secret agent, Henrik was clearly a dealmaker.

"So, here is the deal, boys," he said with finality. "I know some really good programmers. You can pitch them your project and hear them out. If they like it, they'll go for it—given the right incentive, that is." At that, our ears perked up.

"On my shelf," he continued, "I happen to have an orphan like the ones I mentioned earlier that you can adopt. You'd have to work on her quite a bit, but with some good parenting, she might turn out to be a good kid." As both Magnus and I looked at him blankly, he rephrased in plain English.

"I can give you a code base.[5] It's obviously not exactly what you need, but it'll give you a head start. The next step is to talk to David and Frida, and if it flies for all parties, we can put something down on paper. How does that sound, boys?"

In a way, it had never sounded better (perhaps besides the fact that he was calling us "boys" a bit too often for my taste). In a single meeting, we had gone from barely imagining a future for ChallengesЯus to being offered a springboard with the potential of bringing us not only the missing competencies we desperately needed but also a platform to get us started. It felt damn good!

Henrik also wanted in on the founding team, of course —he saw us as a diamond in the rough, and he had his prize for facilitating the transaction. Magnus and I took five minutes outside to discuss our options.[6]

"I'm not sure about this," Magnus said uneasily. "We barely know this guy." He had a point, but I was more willing to jump.

"Even if he's a bit unorthodox, Henrik has to be good," I reasoned. "Otherwise, how would he have gotten this far?" Magnus considered my point. It was clear Henrik knew the ins and outs of entrepreneurship, and he seemed well connected. There was not much space for hesitation; either we went with Henrik, or we had nothing at all.

On that evening of February 17, 2012, everything changed. The three of us shook hands and agreed to a next meeting after talking to the programmers. Henrik committed to speaking with them to explain the base platform on which they would have to build code to turn our idea into a reality.

Not long after, Magnus and I sat down with David and Frida, the programmers Henrik had promised us, and got to know each other.

Though it was practically a non-meeting, we had gotten a sense of their personalities and were slightly taken aback by how young they were compared to us (which soon led to their nickname, "the kids"). David was more analytical while Frida, who was also his girlfriend, was just as skilled but a bit more social. After hearing the details of our project, they were just as excited as we were to move forward. The dream team was in the making, and over the next few months, we registered a limited liability company and started the hard coding of challengesЯus.com.

The kids, Henrik, Magnus and I were now officially shareholders. Although Magnus and I had had to give away shares to make it happen, it was really a no-brainer. Without some similar arrangement, we would not have gone anywhere, and we both knew it.

Despite our compromises, our original pact still held true: we would strive to remain in control of the company at all costs; that is, to retain majority both in shares and in voting rights (we were clearly not educated in law, but we liked to pretend we were—Magnus particularly). It was important.

5. UNINFORMED OPTIMISM

Soon, the kids were working day and night on building our platform. Henrik was right: they were talented. The two of them had their own "web shop" called RedOnion, where they built custom websites and applications for small Swedish companies.

Frida was very quiet, though she had already been awarded multiple design prizes. While David worked on the back end, Frida deployed her talents on front-end development, namely the design language and aesthetics.

Thanks to their hard work, by September 2012, we were ready for beta testing. The idea was to get early feedback from people who were totally unbiased. It was important because if we asked friends what they thought, the majority would just tell us it was great. Beta testers, on the other hand, would tell us exactly how it really was—not with words but through their actions.

The other purpose of beta was to get an understanding of people's online behaviors: where would they click, where would they get stuck, when would they give up, how long would they stay on each page and so forth. Based on those insights, we would tweak our product—not based on our hunches but based on real, demonstrated data.

With some helpful user feedback from the beta testers, we had a successful Skype-assisted launch day—and by February 12, 2013, about 10 days later, our coveted "hockey stick"[1] growth was unfolding before our eyes. It was still early to claim victory, but our user-acquisition trajectory looked phenomenal.

Our ratios were sickening: for every newly acquired user, we modeled that on average, five additional users would automatically be generated. Those five would, in turn, themselves invite five of their friends, and so on and so forth. Soon, our acquisition curve went through the roof. We were gaining hundreds of users per day, completely organically. We had no paid media whatsoever—just our own homemade propaganda.

Invitations to take on challenges were being sent left and right, and challenges were being accepted across the globe. In fact, the number-one metric we had been obsessing over for months—namely, the number of challenges being created on the platform—was showing no sign of slowing down.

With our wide platform finally live, we applied to Sting a second time with all the hope we could muster.

After numerous meetings with Peo and the whole advisory team, we were eventually accepted in October 2012, one of only a lucky 30 or so who made it. Although it was a year later than we had originally planned, it was still a big win.

It had been our indefatigable perseverance that got us in. Magnus and I had been juggling two lives for quite a while as conventional management consultants during the day and wannabe revolutionaries after sunset. Our success had nothing to do with our computer literacy—after all, we mostly just listened to whatever David and Frida had to say on the tech end of things.

We were really an atypical team, and we were raising the average age of onboarded entrepreneurs in Sting by a decade or more. In our young cohort of founders wearing blue jeans with spacers, hoodies and cool hairdos, we earned our nicknames as "the suits."

* * *

There were a few reasons why Sting was the high-water mark of quality within the European startup community. Beyond world-class mentoring, the pure energy that filled the place was second to none. In the Sting offices' huge floor plan, one could find working stations, meeting rooms, cozy "conversational spaces" with fancy sofas and even gaming rooms. It had table tennis and colorful, inflatable beds you would typically find on a sunny beach in the summer. All of it was strategically placed to ensure a natural balance of work and play.

Sting's entire concept was to support a few select candidates by accelerating their time to market,[2] thereby hopefully shortening their path to success. To do that, Sting used numerous support mechanisms such as networking events, business model competitions, pitching battles, conferences, invited speakers, mingling events and even introductions to early investors.

Sting was a more-or-less guaranteed way to secure public money from so-called innovation funds either as hard cash, grants, convertibles or loans. There were always lots of business angels hovering around Sting, vying for a sneak peek at tomorrow's contenders for industry domination. Still, because Sting granted its participants so many privileges, the bar for achievement was also quite high.

Every project team that was accepted into Sting's first stage had to pitch themselves to the committee again in six months to remain in the incubator. A total of three pitches were needed to complete the full program successfully, and in view of the number of applicants, competition was fierce —and our clock had already started ticking.

Unfortunately, even our early hunt for "free money" was not looking promising—we were collecting rejection after rejection. After numerous business mixers and events, we had heard it all: we were not enough of this, we were too much of that; we were "kissing a lot of frogs";[3] our innovative step[4] was too low; there was no market for our stuff; we were too hard to define; we were clearly a US-based venture, not a European one; we had no clear competitive advantage and no means of making money anytime soon; the team was not strong enough; we were too early; we were too late; we were too old (one of my favorites).

We heard a new reason for our shortcomings from every person we met. It was depressing at first, but it became almost laughable as time went on.

Eventually, we met a man named Harry who was a fund manager working with the Swedish government. He was of American descent but was otherwise completely Swedish, and his job was to listen to people like us and decide whether to grant them some small stack of cash.

I had always been suspicious of freebies, particularly when they emanated from some governmental agency whose job it was to support high-risk private initiatives. There was no such thing as "free money"; it was always some sort of options for grant[5] deal. The strategy was designed so that the fund would largely recover any money it granted out, thus making up for any losses it incurred from less successful projects.

Peo had insisted that we meet Harry as part of due course. One hour spent for some 200,000 Swedish Kroner[6] was surely the best hourly rate imaginable, I reckoned, so we went to meet Harry.

We gave our entire presentation to Harry, and after 40 minutes went by, he had not asked us a single question—

nothing about our target market, nothing about our model or our projections. Zilch. Nada. It was worse than that; the guy was literally falling asleep!

As Magnus and I walked away from the meeting, we shared our mutual frustrations.

"Well, that was totally uninspiring," I said. Magnus nodded.

"I don't know how a guy like that can even keep his job!" he replied.

The strangest thing was that two weeks later, Peo came to alert us of a novel achievement: we were officially the first company he knew of in the history of Sting that had ever failed to secure a grant. While we were not happy with our achievement, Peo was furious.

"This is not happening!" Peo shouted. I was seeing the hungry, pushy entrepreneur in him stirring back to life. "I will not have any of *my* companies fail here—ever!"

Peo was not upset with us; from his perspective, we were doing everything correctly, and our project was sound. We were still on the hook for external investments that we'd have to get on our own, but the internal grants were supposed to be a layup. Not getting one didn't just reflect badly on us; it meant that Peo would be the entrepreneurial coach responsible for Sting's first-ever unfunded startup.

After a deluge of emails between Peo and Harry, a month later we were invited to a conference call. This time, Harry was much sharper. He had questions for us that we had no difficulty answering, and the result was that we secured our grant. The return on our time spent was not what we had estimated, but it did not matter anymore. Peo had fought for us, and we had won.

We earmarked the entire Harry grant for platform development. In effect, it meant that all the money would go to the kids to work on the tech platform; meanwhile, the rest of the company (more pointedly, Magnus and I) were still as strapped for cash as ever.

To keep our spirits up, Magnus and I tried to find hopeful parallels between ourselves and Jeff Bezos. It was true, many people laughed at Bezos when he dreamed of eclipsing bookstores with his snail-mail book business—but we remembered fondly that Bezos himself had had to stuff books into envelopes in his living room before Amazon grew into the success story it is today.

ChallengesЯus.com was quickly moving to the top of my priority list. Before getting into Sting, we had thought of our little project as a side gig. But now that ChallengesЯus had successfully launched and was growing, it was getting harder for Magnus and me to stay motivated for our real jobs.

Though the world had recently come out of the Great Recession of 2008, my paycheck seemed to say otherwise. Cole+Moore's customers were not buying complex change initiatives for their companies; they were still being cautious with their investments. It meant I was getting very few commissions from sales—of course, quitting my job would make my cash flow even tighter.

If I left my job, we, as a family, would have to rely on just Claire's salary to make ends meet—an idea she was understandably not too thrilled about.

Claire was a program officer at Sida, the Swedish International Development Cooperation Agency. In Sweden, one percent of the country's gross national income is used every year to support development abroad in poor countries, and Sida's staff manages those funds and uses them to pay for programs to combat global poverty. Like most Swedish bureaucrats, her paycheck was not grand.

Though our options were limited, we did have some savings, and I estimated that our family runway—the time Claire, Chléa, Melvin and I had before running out of cash—was between eight to 12 months. Though it was a risk, the company was gaining momentum, and I needed to make it my main focus.

"Our savings will give us ample time to complete our first funding round," I told Claire with confidence. Inside, I prayed that my optimistic predictions would come true. After getting Claire's support on a Friday night, Magnus and I set up meetings with our respective bosses that coming Monday. Mine was with the managing director of Cole+Moore Europe.

* * *

David was an engineer from Umeå, which sits on the Gulf of Bothnia in Sweden, on the northernmost stretch of the

Baltic Sea.[7] When we started working together, he had been out of college at the University of Umeå for less than ten years.

As I'd learned in the brief time we'd known each other, David was rather terse in the way he expressed himself—down to his habit of saying "yes" as a simple slurp or inhale.[8]

Perhaps it was part of his culture, his lack of confidence in English or all the above, but while we were preparing for launch, I sometimes wondered if I was the cause. Was I stressing him out? Not purposely, of course, but just by being so unlike him in so many ways?

Most of what David said were things I couldn't understand. He mostly spoke in PHP,[9] with occasional references to AJAX.[10] To me, AJAX was only a Dutch professional football club based in Amsterdam. Either way, there was never any hyperbole in David's language; he was a programmer after all, not a lyricist.

And so, as we approached the milestone of 8,000 users that Friday night, he was quick to raise a red flag.

"I'm not sure how many more users the platform will be able to handle," David said out of the blue.

"What do you mean, man?" I asked immediately.

"Well, we built the platform on top of something we inherited," he explained. "I did not write that initial code myself. I don't know if it's any good. Of course, I patched the obvious things as we moved along, but it's been quite hectic until now, I must say." It was not a very satisfying answer.

"Okay, David," Magnus began, "I think I understand what you're saying, but can you be a bit more specific?"

"Basically, it could crash anytime," he replied.

"*What*?" Magnus and I squawked in perfect synchronization. Frida jumped in immediately, sensing that both Magnus and I needed slightly more reassurance. She tended to intervene in situations like these, thankfully—she was good at reading people, and in her company with David, she was the one who dealt with customers.

"There are a lot of things we don't have in place," she said. "You have to understand that when people start doing challenges and upload their photos and videos—particularly the videos—and then people from all over start participating with likes and comments, there are a *lot* of database requests to handle. That's a heavy load on the platform, especially if—or when—a lot of people do it at the same time." What Frida was saying was enlightening, though still not very reassuring.

"We have not done any of this before," she continued, "so it's hard to say what will actually happen. On top of that, we clearly don't have the capacity to host large quantities of media files at the moment. We rent part of a rather small server if you remember; it's not like we own racks and racks of them."

I did not like what I was hearing, but it was now crystal clear that we were in a bit of a pickle, to say the least.

"What do we do now?" I asked. "Guys, any ideas?"

For the next few hours, we contemplated our options. We stayed up all night and spent the entire following day at Sting, taking turns stretching our legs when we could no longer sit. By 1 am, the common fridge, which we'd been raiding since 6 pm, was nearly empty. We would have to wait for the suppliers to come and fill it up in the morning—that is, if they even worked on Saturdays.

We all felt like we were on a long-haul flight, with a similar set of bodily sensations including exhaustion, stress and anxiety. During that time, we kept an eye on the physical counter we had purchased on eBay to keep track of our user count. It was a very cool gizmo with a sleek design, a bit like a retro clock. It kept climbing.

By the time Peo showed up around about 9:30 am, we were shattered. Magnus was taking a power nap while sitting upright in his chair with his head lolled back and mouth half-open. He had mastered the art of falling asleep anywhere, anytime, a skill he had acquired after spending so much time on planes on assignment for Cole+Moore.

The kids had their heads down and were fixated on their screens, hammering madly at their keyboards. Between the two of them were three laptops. David had two and was swiftly wheeling his office chair on the floor from left to right to left again. Despite his sleep deprivation, his footwork still reminded me of a professional tennis player. As for me, I was downing cup after cup of diluted black coffee. In between, I would drop to the floor and do 30 to 40 push-ups at a time to stay awake.

We explained the tech situation to Peo, who was also perplexed. There was not much we could do now; migrating everything to a bigger server would take a few days at least but likely longer—and the last thing we wanted to do during a period of high growth was to interrupt our service.

The only quick fix we could come up with was to cap the size of the content people would be allowed to upload. While this sounded like a reasonable proposition, it meant that people trying to upload videos that exceeded the set size limit would now not be able to do so—and they would not realize it until the long uploading process was over and ended in failure (unless they read the fine print by the upload button, which no one does).

What would our users do then? Even if they were to understand that there was a file size limit, would they go through the extra work of downsizing their videos? Or would they just quit? I knew what I would do. The four of us mostly agreed that we would just give up in the same situation, but we concluded that perhaps we were not a representative enough sample. Regardless of our opinions, the kids put the solution in place that night; it was the only thing we could do at that stage. Unfortunately, there were more hiccups to deal with.

It must have been about 3 am when a challenge caught Magnus's attention. Christman99 from Tampa, Florida,

USA was challenging HarleyJeff, also from Tampa, to get into the lion enclosure at the local zoo. He had one week to complete the challenge and upload his proof. At first, we laughed—but then we panicked.

We had not made any provisions for reckless challenges. Our users' Terms and Conditions were rather thin, and we had written them ourselves, just like our shareholder agreements. It was how we had drawn up all our official documents.

We thought we were being good entrepreneurs by jumping first and building the plane on the way down with our Terms and Conditions, but it had serious implications for our business—and possibly for us as owners. Compared to our users being eaten by lions, working on tech scalability now seemed like a luxury problem.

It was fortunate that our small team got along so well. Though it was the middle of the night, and we were all clearly exhausted, we quickly divided the work. Magnus went on to review all our so-called legal documents while I went through the platform to scan for hazardous challenges like the one we found in Tampa. In addition to the quick fix to limit file sizes, Napster and Frida jotted down ideas on what they could build to proactively stop harmful challenges from going live on challengesЯus.com.

There were a lot of challenges to vet. Beyond the benign ones like "learn to play the guitar," "register for a marathon" or "shave your head tonight," there were plenty that seemed like they came directly out of *Jackass*. Some challenged people to ride *on* buses rather than *in* them. Others were about drinking dangerous amounts of alcohol or drinking dangerous amounts of water—both of which could lead to irreversible brain damage, seizures, comas and even death.

There was even an attractive girl from Bratislava, Slovakia who offered a tête-à-tête dinner to the first person who would run across the tarmac of any major European airport. What was crazy was that it was impossible to say how many foolish challenges had been created so far. Now, I had to go through them all one by one—hundreds of them.

We concluded that the best way to deal with this issue was to add an algorithm to the back end of the software to look for red-flag words like "drinking," "drugs" or "lions." It also made sense to offer our users a flagging option to report on abusive and dangerous challenges.

Thanks to Peo, the legal topic was quickly under control; Sting had direct access to the best lawyers in town. By 2 pm, even though it was a Saturday, we had one of Sting's legal partners turning our layman's legal mumbo-jumbo into bulletproof documentation that would clear us as private individuals from any lawsuits our company might face as a result of accidents, even lethal ones, caused by the use of our platform.

We had narrowly dodged one bullet, but there was no time for celebration. We still had to address our core tech issue, and for that, we needed more money. We would also soon need to pitch ourselves again to Sting, and there was absolutely no guarantee that we would be able to remain in the program. To top it all off, I had a meeting on Monday to tell my boss that I was quitting my job—and I was starting to second guess that decision.

6. A LIFELINE FROM CAROLINE

My boss, Caroline, was seated at her desk when I showed up at our agreed time. She had an open-door policy at work that was quite literal, so I stepped in the way I normally would when I had something to discuss with her.

"Hey Caroline! How are things?" I said brightly.

"We'll see after this," was her short answer. Caroline was not one for small talk. She had made her entire career at Cole+Moore, moving from junior consultant in her early 30s all the way to the top in just over 15 years. That was quite impressive, perhaps even more so because the consulting industry was often a playground for privileged, white, middle-aged men. Caroline had been adopted from South Korea by a Swedish couple at the age of three, a background that had almost certainly contributed to her resilience.

Caroline was a strong leader, and she would not hesitate

to speak up whenever someone unwisely implied they were tired or could do with a raise. Some in the office thought she was ruthless, but I disagreed; she just had no sympathy for slackers, which was a good thing.

In truth, I enjoyed her management style. We'd had our arguments, but I didn't blame her for them. She had what we would define today as a "growth mindset," albeit taken to the extreme. Magnus did not necessarily buy into her philosophy, and according to him, she knew it.

"What can I do for you, Arnaud?" Caroline said. I took a moment of silence to collect some strength.

"Well, Caroline," I began, "I, uh, well…I quit."

"I know, Arnaud," she quickly responded. "I know." It was not the reaction I had been expecting.

"What do you mean you know?" I asked, rather surprised.

"Well, I am not a psychic, but I've noticed that lately, you have not been as passionate as you used to be."

"Oh, well, that's…good," I tried.

She raised her head above her laptop screen, behind which she had been hidden since I entered the room. She was now looking straight into my eyes. It almost felt like she was trying to fry my brain.

"I mean, it's good that you know," I tried again, the situation only getting more awkward by the second. She exhaled profusely through her nose.

"Okay, so tell me," she said. "You got an offer you

could not refuse? Is that what it is about?" For a moment, as I was looking for something clever to say, I thought about Magnus. Caroline didn't know that at the very same instant, Magnus was having a similar conversation with his own direct supervisor.

Magnus and I were in the prisoner's dilemma.[1] Of course, Magnus and I were aligned on the fact that we needed to quit. The time was right, we had passed the limit of what we could do for Cole+Moore, and evenings and occasional weekends were no longer sufficient if we were to succeed on our own.

In our case, the two of us acting in tandem pursuing our common self-interest would produce the most optimal outcome for ChallengesЯus—but still, what would happen if during their conversation, Magnus's boss started dangling the promise of a massive salary increase in front of his nose? I didn't think he would chicken out and accept a raise, but now the thought was crossing my mind.

"No, it's not another offer," I replied, snapping back to reality. "I am launching my own company." Once again, Caroline's reaction was swift.

"You're what?" She asked, laughing like I had cracked the funniest joke she'd ever heard. "You're leaving Cole+Moore for the dream of—of what exactly?" I didn't know what to say, so I started nervously laughing myself.

"Look, Arnaud," Caroline began, "you and I have had our differences, but you know I like you, right? I think you are a nice guy. You are also a good sales guy—a great sales guy, even. So, before you do or say anything stupid, I just want you to know that you can take back all this nonsense. I will forget all about this conversation, and we will go back to business as usual, just as if nothing had—"

"It's already live," I said, cutting her off before she could complete her sentence. She looked like she had seen a ghost. I could feel her assessing whether I was sane. Soon, the conversation changed from an intimidating talk between boss and submissive employee to a pleasant chat between two professionals. I started explaining my project to her—and to my complete astonishment, she was buying into it.

She was not just intrigued by the idea of me starting a company in an economic downturn but also the actual concept and what the company aimed to do. She eventually confessed that it could very well make sense, which was not something I had expected.

Caroline had two boys around 14 and 15 who were always glued to their phones. She explained that ChallengesЯus might be useful for her own anti-social offspring—and probably most of their friends, too.

"They would spend a lot of time on your platform," she said. "This has the potential to be at least as addictive as Facebook. I'm actually starting to like what I'm hearing, Arnaud."

She went on to ask very specific questions about the

model, how it had been financed and how we intended to monetize. Her business inclinations were taking over, and the whole exchange felt surreal. I had come to quit my job and was shit-scared about the prospect of breaking the news. Now, here was Caroline, interested in my idea. Was this really happening, or had my mind gotten ahead of me?

Caroline and I went on to have an hour-long conversation. To my absolute astonishment, she eventually asked, "Can I join in?"

She's pulling my leg, I thought. Was I being framed somehow? I started to wonder whether it had been smart to tell her everything I had just told her.

"I'm serious, Arnaud," Caroline said. "I like the idea. I mean, obviously, I am not going to quit my job tomorrow for this, but I'd be interested in investing some money if we could agree on something that made sense."

After a little more discussion, we agreed: I would get back to her with a slide deck and an investment pitch. In my excitement, I felt compelled to invite her to check the service out for herself.

"Why don't you challenge a few friends of yours and perhaps even yourself to something?" I suggested with excitement.

"Yeah, yeah, I'll see what I can do," she replied firmly while plunging back into whatever she was doing before I entered her office.

As I walked back to my cubicle, I could hear Magnus's voice echoing in my prefrontal cortex. He had a signature saying he'd utter as soon as he encountered any situation that didn't fit with his *weltanschauung*, [2] which was often. He would say, "If it's too good to be true, it probably is." In this case, I prayed he was wrong.

As I sat down at my desk, it took me a good 10 minutes to get back to my normal state of being. I reached for my phone and called Magnus. His phone number extension was 555, which I had joked about in the past. In numerology, the number "555" represents what you're going through in life and positive changes to come. Pressing that number in at that very moment felt like the sky was opening above me, and the stars were aligning for Magnus and me.

Magnus's office was right below mine, so I could hear his phone ring, but he was not picking up. I tried his cell, but to my surprise, he had already left the office and was now on his way home.

"The meeting was over in less than 20 minutes," he later confessed to me. As it turned out, his manager had been very supportive as well.

After officially resigning, we quickly moved on to ironing out the deal details and formalizing a partnership. Just like that, Caroline, the Iron Lady, managing director of Cole+Moore Europe, became the newest shareholder of ChallengesЯus. In the next few days, she deposited €70,000[3] in our bank account as our first business angel. We had made it to startup heaven.

I thought of a passage from the King James Bible:

"Ask, and it shall be given you; seek, and ye shall find; knock, and it shall be opened unto you." We had needed cash to operate, and it had manifested itself almost effortlessly. This was a direct sign sent from above—no doubt about it.

7. GROWING PAINS

With the money we'd raised from Caroline, ChallengesЯus moved into its own office in Kista, a suburb of Stockholm. Kista is the largest ICT cluster in Europe and the world's second largest after Silicon Valley. The Ericsson Group, the largest corporation in Sweden, but also IBM, Oracle, Samsung, Citrix and many other companies have offices there. It is also the research park of KTH, which is why Sting had space there as well.

Unlike the space Sting had downtown, our space in Kista was not a co-working space but an actual office.[1] It was a shared office, of course, but for the rent we paid, it was a good deal. We shared the conference room, the coffee corner and a receptionist with two other companies. On our door, we put up a large challengesЯus.com sticker.

Upon entering, you arrived straight into what we called "the oval office"—not because it was white or oval-shaped but because we had arranged our desks in an oval formation so we could all face each other.

I had also taken the time to bring some of my favorite business books to the office for decoration,[2] though I'm not sure why—there was no time whatsoever to recline with a foamy mocha while flipping through a book I had already read. Cole+Moore had had plenty of books, so it felt like a must—every office needed books, right?

We spray-painted our tagline onto the biggest wall in the room and had an artist friend come in to paint us a nice mural. We also brought in a few plants to please Frida. Initially, the plants felt more like a "nice to have" than a "must-have," but I was glad we had brought them. It made the space feel like home.

* * *

Several months past our official launch, users had sent hundreds of challenge requests from our platform all around the world, roughly one in three of which were accepted. We had nothing to benchmark against since no one else was doing anything close to what we were attempting to do, but a 30 percent conversion rate, give or take, felt like a good starting point. Still, an accepted challenge was not the same as a completed one.

To validate our concept, we needed completed challenges. Without completed challenges for users to enjoy and be inspired by, challengesЯus.com would be like

YouTube without videos. To our bewilderment, most of the challenges posted had already expired. Worse, very few of them had been completed—even the people who had challenged *themselves* had not managed to live up to their own aspirations. The cherry on top was that not a single new challenge had been created in over a week.

We started asking ourselves scary questions. Were people really the social animals we had imagined they could be? Could they be more than couch potatoes who only shared ridiculous pictures of cats wearing party hats? Would they go beyond their comfort zones and into the real world?

Our mechanics for viral spread did not seem to be working as we had envisioned after all, and we had somehow come to a standstill. It looked like all our enthusiastic users had vanished. How had it happened? Was the world under nuclear attack? What about the youngsters and their *Jackass*-style challenges? Where had they gone? Surely not Facebook and definitely not Pinterest. Maybe some other company had launched a similar application at the same time as us?

As we turned to social media to investigate, it was the same story: dead quiet. How could we have gone from global euphoria to absolute silence so quickly? There had to be an explanation—a rational one, we hoped.

Since launch day, each of us had been laser-focused on our tasks. Magnus was knee-deep in paperwork, dealing with everything from navigating the intricate process of onboarding new shareholders to filling out official documents for the tax authorities, auditors, investors, banks and so on.

On my side, I was multiplying challengesЯus.com's meetings with advertisers, showing off our platform and selling them on how their brand would soon be able to target people according to demographics and type of challenges. It was an easy pitch compared to the "strategy alignment" pitches I was used to.

For every phone call I made to ICA,[3] VisitSweden[4] or even Red Bull, I would have their executives excited in no time. They all wanted to be associated with an active, fun and young demographic. The potential was unlimited, and advertisers' appetites seemed huge.

Magnus and I were also working with Peo on our growth strategy and started screening candidates for both sales and tech roles. For their part, the kids were coding as if there were no tomorrow. Their key priority was to enable the platform to scale. For starters, they would have to "clean the code," though we were also buying more server capacity. Because we had all been so busy, none of us were paying much attention to what was happening on the platform itself. We had put all our trust in our concept, which was a rookie mistake.

We were in shock when we realized our engine for growth had stalled—and Caroline was far from happy. She had just deposited her hard-earned cash into our account to help us fuel the supersonic growth in user acquisition she had witnessed earlier, but now that momentum was gone.

Soon enough, Caroline was calling me every other day for updates, but I did not have much to report. We were just as confused as to why everything was collapsing.

What could I tell her? *Well, Caroline, I guess we fucked up somehow. Sometimes you win, sometimes you lose. Tough shit—sorry!* No simple aphorism would solve the situation, and Caroline wouldn't abide any excuse I came up with, even on my most imaginative day. Although everyone knows there is a high degree of uncertainty in any startup venture, Caroline was not very comfortable playing Russian Roulette. To keep her happy, we needed to fix our growth problems and fast.

It was dawning on us that Zuck would not be calling anytime soon to ask our price. More work would have to be done, as well as more blood and tears poured in. There was simply no other way around it: we had to get in the trenches with our now-moribund platform.

The first thing that came to mind was that the tech was sub-optimal; it had been built in a hurry and out of the corpse of another project. Something had to be buggy! That's why no challenges had been completed or created! Software had to be the explanation, we reasoned, because clearly our concept was flawless—we had proof of that since people from around the world had flocked to it like seagulls to hot chips.

This idea to challenge oneself to improve, as well as to

challenge one's friends for a bit of fun, seemed universal. It transcended sex, age, culture and ethnicity. Even though people were creatures of habit, we were also constantly working at (or at least dreaming of) bettering ourselves. It was all we had ever done since the dawn of time. We explored lands. We made scientific discoveries. We searched for new things to do and invent. We learned, always. It was just what human beings did.

It's what Abraham Maslow called self-actualization in his famed Hierarchy of Needs.[5] It's how we progressed from caves to campfires to building empires with gold and silver. In so many ways, we had not invented anything at all with our challenge concept; we were merely putting something humans were already doing in the real world online.

We decided that we would allocate all our time and effort to improving our product—its interface, the sign-up process, challenge onboarding, content uploads, content sharing on Facebook and other platforms and so on. As we de-cluttered the code and migrated to bigger, more powerful servers, we increased the file-size upload limit—of course, that did not help at all since no one was creating challenges. Even so, we were getting ready for our comeback.

We "dumbed down" the platform as much as possible by adding big fat buttons people would not be able to miss. We removed all superfluous text, and we made it possible to challenge directly from Facebook—there was no need to even have an account with challengesЯus.com to get started anymore. We had our own toddler-granny acid test. Basically, the platform had to be so easy to operate that both a toddler and a grandma could sign up and challenge someone without a tantrum from the former or confusion from the latter.

We A/B-tested different features. We ran analyses. We worked hard to improve our ranking on Alexa, which assesses how a website is doing for traffic relative to all other sites on the web. We also went on the marketing offensive, looking for leverage anywhere we could.

We compared the click-through rate of different ads on Facebook to see which ones resonated most with our audience. We pushed to get interviewed in local newspapers. We printed leaflets and hired students to distribute them at universities and beyond. Whenever the team was out walking, shopping and relaxing with friends and family, we used every possible opportunity to market the hell out of ChallengesЯus. We left leaflets on car windshields, in bars and on the train—anywhere and everywhere.

We had what we felt was an awesome tagline: "If it's not on challengesЯus.com, it probably never happened." We would yell it in public spaces, and we loved seeing people's reactions when we did it on the metro. We got our tagline embroidered onto flashy orange hoodies that we wore as often as possible.

We ordered hundreds of black caps from China with a big orange C on the front with the plan to send them as prizes to winners of challenges we would sponsor. We threw real-world live challenges to raise awareness—and of course, we participated in our own challenges to lead by example.

Our small office was buzzing with activity, but despite our efforts, nothing was changing. We dug deeper, calling on our friends to hear their opinions. Why were they not creating challenges? Why would they not complete our own challenges? We heard all sorts of reasons: they couldn't find the time, they had forgotten about it, the things they were challenged with didn't feel right, they got sick before the end of their allotted time and so on.

The more we learned, the more confused we got. Some told us we were clearly lacking a mobile app, which was easy to say but harder to conjure up. Developing a native mobile app could easily mean spending millions; it was easy to dismiss since we did not have that kind of money. There were so many other things to consider before coding an actual native app. For starters, it would mean learning a whole new coding language, and David and Frida were not warm to the idea of spreading themselves even thinner by working two different platforms at the same time.

The main difficulty was that we would have to match every little change we made on both platforms, which would slow us down dramatically. We were trying to grab the market before someone else with deeper pockets, most likely in Silicon Valley, ran us over. We had to find shortcuts.

To get around making a native mobile app, we made

our platform "responsive," which meant you could access it on a mobile phone via the browser. It didn't offer the same ease of use as an app experience, but it was a start. Unfortunately, tackling the problem this way had limitations once you were on the website.

For example, it was difficult to use the platform's buttons because they were so small on a phone; users would have to zoom in and fiddle around with drop downs. It was also quite painful to tag challenges or even write a description for challenges. On top of all that, our browser-based solution was not connected to the phone's contacts directory, so trying to invite people was a horrendous process—and kudos to anyone who even made it that far! As much as I wanted to be optimistic, it all seemed like a lot to ask.

As far as the mobile app was concerned, our conclusion was simple: it had only been two years since Facebook released its own Android app, after eight years in business and billions in revenue. Surely, no one in their right mind would come to expect the same experience coming out of a newly launched company. Or would they?

* * *

To raise office morale, someone suggested that we get an office mascot—though what kind of mascot remained an open question. An office dog was ruled out quickly; none of us had one at home to begin with, so getting one for the company seemed excessive. The kids had a chinchilla at home, but we did not quite feel like bringing it in the office. We finally decided on another type of mascot: a garden gnome.

A friend of mine had given me the gnome as a joke some years back, when he visited from France when Chléa was born. Claire hated the idea of having a garden gnome in the house, as it is so completely not a Scandinavian thing to have—it wasn't even particularly welcome outside. The result was that the poor bastard had been hiding in the basement of our house ever since. That is, until I rescued him, painted him orange and stood him on our office's only cabinet.

We named him Charlie and even gave him a spot on our website, under the team tab—all of which, we thought, added to our quirkiness and helped us stand out. Still, Charlie was more than just a mascot; he soon became a part of our marketing strategy.

In *Amélie de Montmartre*, the massive French hit movie from 2001, the main character, Amélie, steals her father's garden gnome and gives it to her flight attendant friend. Her friend then sends her dad photos of the gnome traveling abroad. It visits London, New York, the Cambodian temples and many other places. Similarly, we put our orange gnome in various situations and captured those scenes as well; we filmed short movies and released them strategically to generate interest.

The hope was to start a movement that would see ChallengesЯus spread on social media like wildfire—and it did, at least a little bit. People were sharing their own garden gnomes in all sorts of situations and locations. We saw bungee-jumping gnomes, diving gnomes, climbing gnomes, tanning gnomes, dancing gnomes, goal-keeping gnomes and more. We even came across this Russian gnome flying a Sukhoi SU-27, which is a twin-engine super-maneuverable fighter aircraft.

Thanks to our little mascot, we came to realize a critical missing link in our business: people needed a little help to keep going. This was not a big *aha* moment of course; one could easily see that to be true everywhere in the real world. Personal trainers helped people achieve their fitness goals, managers kept pushing their teams, parents guided their kids and so on. It was everywhere.

Magnus and I had often debated human traits, mental models and what made people so successful. It sprang from what we had studied, and what we had learned and taught executives at Cole+Moore. We were passionate about the "people factor" of extreme performance and achievement, not to mention discussions of mental strength versus physical aptitude, nature versus nurture, genetics versus environment and so on. There was no end to it.

Of course, at the core of our exchanges was always how we would leverage those human principles to build a better challengesЯus.com. How could we bake it into our platform? How could we cascade it into our features? When we met people and explained our vision, we could get excited about what was under the hood—all those mechanics we had built and why—but people in general didn't care that much. They wanted the short version, the bullet points.

Finally, it hit us: what if we were to offer ready-to-go challenges? People could sign up to them with one click, a bit like Amazon's one-click purchases, a huge breakthrough when it came out in 1999. If we could emulate that idea and turn online challenges into a hassle-free thing to do, we could help people do things they intrinsically wanted to do but so often didn't. Couldn't that be something? Even if we still had the issue of challenge completion, we could at least dramatically lower the barrier to entry.

Magnus and I worked like crazy for a couple of days on our concept before presenting it to the kids for feedback. Frida and David never really shot down any of our ideas; they shared their points of view, mostly from a technical perspective, and let us decide what we wanted to prioritize over the million other things we had on our developmental to-do list. In the case of our ready-made challenges though, they were enthusiastic.

As a result, we discussed it together and extended the concept even further—until I finally had an idea of how we would get the traction we so desperately needed.

"Maybe people would be inspired by celebrity involvement?" I said to the team one day.

Soon, everyone rose in a chorus of approval that sounded almost like when British MPs voted on a piece of legislation and welcomed it with a rising *aye*!

"That's an awesome idea, Arnaud!" said Magnus.

"Yes! Dorsey, here we come, baby!" I said, thrusting my fist in the air and feeling victorious. I was referring to Jack Dorsey, of course, co-founder and CEO of Twitter. I never quite understood Twitter (I often wonder if anyone does), but one of its applications is that it allows users to "follow" each other.

As it turned out, people tended to follow famous and influential people. The more we thought about it, the more obvious it seemed that engaging only with tweets might soon be a relic of the past. SMS might have been cool when the social web was in its infancy, but would that be the case for years to come? Would it not be more rewarding for both parties, the follower and the celebrity, if there could be more of an exchange?

Of course, no one could expect LeBron James to have one-on-one chats with his followers; it would be nearly impossible in the real world. So, the question we asked ourselves was: what could ChallengesЯus propose that would help celebrities like LeBron promote themselves and their brand while engaging their fans? In LeBron's case, what if he were to challenge his followers to a slam-dunk contest, for instance, and share it? Would that not be a compelling proposition?

Or what if the English comedian Sacha Baron Cohen challenged his followers to prank people in the streets as Borat? Wouldn't his fans be more engaged by that than by reading his tweets? Wouldn't it make those celebrity brands explode even further? Wouldn't it make the world a better (or at least more fun) place?

Armed with Internet Explorer and a phone, I went on a mission to corner the rich and famous, the Shaquille O'Neals and the 50 Cents, the André Agassis and David Blaines of the world. Clearly, Jay Leno's mobile phone was not available for anybody to access, so I had to be both strategic and patient. I was shooting out emails and Facebook friend requests like a military-grade assault rifle, but from American surfer Kelly Slater to Czech supermodel Eva Herzigova, very few were getting back to me.

Some were kindly declining the invitation (perhaps not directly but through their social media teams); those replies were nice. Others, though, were downright rude. I had been in sales almost all my professional life, so rude replies did not bother me much anymore. I knew B2B statistics, but I had no competency whatsoever in the industries we were now trying to poke. Still, I was excited to learn.

How many outreach attempts would it take to get a positive answer? That was the question everybody on our team was asking. The answer, as it turned out, was a hell of a lot. Still, I had thick skin; I always viewed canvassing as straightforward arithmetic, so the more no's I received, the closer I was getting to a yes.

Soon, I concluded that convincing the world's top celebs to join us without a proof of concept was going to be more challenging than I had anticipated. Next, Plan B went into motion: going for other-list celebs. I would continue to hammer the A-list guys, but only the Swedish ones; after all, our company was Swedish, so maybe it would help us build trust. At the very least, it couldn't hurt us. If I succeeded, I would even be able to hop on a train and take in-person meetings with celebrities—who wouldn't want to do that?

I started going very "niche" to try to generate some sort of snowball effect. There was a hardcore gangster rap group from Louisiana I had found by following threads on Facebook. They had some cool songs. I had somehow managed to hook up with them virtually.

My guerrilla marketing strategies were miles away from my MBA classes and light-years away from my strategic discussions with executives of listed companies. It was also so not in sync with my vision of building a tech startup. All the same, I kept telling myself that sometimes you had to step way out of your comfort zone and do things no one else was willing to do. It was where all the growth could be found.

I was taking video calls with badass gangster groups in the middle of the night to make sure neither Claire nor my kids overheard the verbal carnage. We were talking about doing a rap that would lift challengesЯus.com and smoke Facebook—it was a long shot for sure, but I had nothing to lose. Still, after many odd conversations, I got tired of their rubbish; they wanted too much money upfront, and I wanted substance and commitment before sending any.

On the Swedish front, things were easier. Celebs'

mobile phone numbers were readily available if you knew where to look, so I was calling all the household names I could find, one by one. Before long, I was having conversations with Magnus Samuelsson[6] and Nicklas Kulti.[7] Even if they were relatively small, it felt like momentum, and I was having the time of my life.

To the team's astonishment, a month into hunting, we inked our first celeb out of the US, and it was none other than Eric Larsen. Eric was a world-renowned explorer whose biggest accomplishment, up to that point, was that he had gone to the South Pole and North Pole as well as to the top of Mount Everest all in the same year. This time, his plan was to bike with his custom-designed Surly Moonlander machine in -30C weather. Nobody had ever traversed the South Pole by bicycle before, and ChallengesЯus.com was to be the platform covering his very special world-record attempt.

The partnership generated a lot of PR, and we used every bit of it. Thanks to his satellite phone, Eric did some livestreams on our platform straight from his tent in the middle of the South Pole. Over a period of 10 days, he invited his followers to do things like get on their bikes as much as possible, along with some touching challenges aimed at raising funds for Parkinson's disease research.

It was grand. We all felt like the good guys, and overnight, our vision magically transformed into something greater than helping people challenge themselves and others. We wanted to do our part to positively impact the world. Our fans loved it, and so did Caroline. With such a great calling card, it suddenly became way easier to get people on the phone.

Very quickly, we were in business with several super achievers in their own respective fields. There was Decorey Jones, an American NBA draft-eligible basketball player, now playing professionally in Thailand. We had Sarah Nyberg, Swedish fitness celebrity, as well as David Persson, world-class illusionist. The list went on.

We almost did the same trick we had done with Larsen with Dogge Doggelito, one of the most acclaimed rap artists in Sweden. Dogge's plan was to be sent to space, and we pitched his managers on having him kick some of his rhymes from space via our platform while he was up there. Had it ever actually materialized, it could have propelled our startup to new heights—literally. Unfortunately, it didn't. I'm not sure why the project was abruptly abandoned. All I know is that Dogge did not go to space, and we didn't either.

Over the next months, we worked with celebs and their managers to create custom contracts with all sorts of nondisclosure clauses—and yet once more, despite the unparalleled inspiration we were supplying, the vast majority of our users were choosing not to participate. Instead, they watched from the sidelines. It was shocking to us at first and then outright demoralizing. In the end, however, it was quite illuminating.

We had found the crux of the problem, and it made all

the sense in the world. It was clear that the vision of an improved future self had led many people to sign up for challengesЯus.com. It was also obvious how those who wanted the best thing for their friends, their parents and their partners had signed up to push their loved ones into making changes they needed to make. And of course, it was clear why the person being challenged was thrilled by the possibility of changing, only to later abandon it for fear of failure and feelings of inadequacy that might follow.

With that, we had found the very reason Facebook was working so well. How could we have missed it before? How could we, the change management experts, possibly have missed what resided at the core of every journey of change?

It was now perfectly obvious why no one would want to commit to something they were already completely convinced they could not achieve—especially when the whole journey was going to unfold online in the public eye. Why would anyone in their right mind volunteer to be potentially ridiculed or lose face on the global stage?

Our guerrilla marketing experiments preceded the Ice Bucket Challenge[8] by over six months, but while that went viral, we did not. Instead, we were already gasping for air.

8. PIVOT

When the talking heads on CNBC have tense discussions about negative growth, people don't necessarily pay attention. But as a startup, when you experience so-called negative traction,[1] you cannot automatically think that things will be fine because there may not be another quarter for you to make it back. Without securing a lifeline, you might even go under.

When that happens, it feels like everyone in your circle is talking about you and your failure. No one wants to touch you with a ten-foot pole. After all, why would anyone want to be associated with you, not to mention invest in you, when you are digging yourself a hole you can't seem to climb out of? It's the same time your friends start telling you that they warned you and that you should've stuck to the cozy job.

It's when your former colleagues are glad they didn't indulge in your folly, when your wife starts seeing the dream of owning a spectacular house by Lake Como, the one you've been promising for years, rapidly evaporate. In our case, even our fellow growth-hackers at Sting, the ones who intimately knew what it took to put a dent in the universe, started looking at us with disdain.

"We are done," Magnus kept repeating. "The party is over."

"It ain't over till the fat lady sings," I was always quick to respond. We might have failed in our first iteration, but I still had a fire in my belly to go in a new direction. We still had the tentative support of Sting and a talented team determined to make the project work. I was not hearing the fat lady's music. Not yet.

Our respective personal runways—the startup measure of how long one could survive on personal savings—were getting shorter by the day, too. Claire and Josefine, Magnus's wife, were also not helping. It's not that they weren't trying—they were. As a matter of fact, it was thanks to them that we made it as far as we did. Everyone knows the saying that "behind every successful man, there is a strong woman." While we could not yet confirm our success, what was certain was that Magnus and I had strong women beside us.

Neither Magnus nor I had received a single penny for nearly a year. All the money from Sweden's innovation fund and from Caroline had gone straight into product development and operating costs, which included paying the kids.

We had a special kind of arrangement with Frida and David. Since they were not ready to drop their own business when we joined forces—which was a very smart move for all of us, by the way—we had them on the payroll but only as consultants. Without getting too deep into the details, this reflected better on our company's statements. And so, to incentivize them to work hard for ChallengesЯus, we had given them shares—a good chunk of them. The real present value of those shares was close to zero, but if we were to make it one day, they would most likely not have to work ever again.

We became very anal with our day-to-day expenses and tried to get a return on investment on everything. We were working constantly, including nights and weekends. Forget taking off early on Fridays (aside from rare special occasions such as events organized by Sting or when we were pitching abroad)—it was just work, work, work. Still, the more we worked, the less clarity we seemed to achieve.

Over the years, I had heard "well-being gurus" propose that there was absolutely no correlation between work and results. In fact, evidence seemed to indicate that the work-produced-to-impact-created ratio followed the classic Pareto principle like everything else: 20 percent of efforts produced 80 percent of the results. By that point, ChallengesЯus had an inverted Pareto. We were trying to find ways to reverse that, but it was hard to break the cycle. Fires had broken out everywhere, and we were seriously starting to wonder whether we had the skills to keep a company afloat at all.

With our little "crisis" showing no signs of stopping, Peo told us that it would be borderline impossible to raise money with no income streams—even if we did manage to generate a lot of free users in a short period of time.

"The time for your kind of business-to-consumer business that grows only on free users is over, I'm afraid," he told us. "Right now, I don't know any VCs in Europe who would want to invest in that kind of model. They've learned their lessons. They've been burned a few times. It takes too much time, too much capital and too few have made it to justify the risk. If I were you guys, I would start thinking of a Plan B."

Initially, we refused to even listen. We had a "we will prove him wrong" mentality. We were entrepreneurs after all, not fifth graders. We were in it for real—'til the bitter end. Peo knew that, of course; it was part of the reason we got into Sting in the first place. We were not going to crawl around like dogs waiting to be put down (not now, anyway). As I told myself that story, I pictured the shelf on Henrik's wall at KTH. I could even hear those demonic crying babies loud and clear. It was excruciating.

As we desperately needed income, I picked up the phone one day out of total frustration and called on a few of my former business contacts, ones I deeply trusted. If we were not going to generate any income in the short term from ads and there were no other viable models available to us, should we not, at least as a short-term fix, instead try to sell challenges to corporations?

After all, why would companies not buy our services for their employees, challenging them to undertake whatever was relevant for the company's internal operations? It was not the plan we had when we started the company, but it could help us generate cash flow while we figured something out.

I had a lot of experience in the B2B trade, and I knew that it typically took months or even years of orchestration with a multitude of internal stakeholders to close big deals. Even so, the heads of HR and employee engagement agreed to a call.

On our first call, I painted a picture of how we could challenge their employees globally and of our culture as a growth-oriented tech company. I tried to ask more questions than I answered since I needed to assess their appetite and learn their current organizational headaches. Of course, despite their apparent enthusiasm, they were not about to sign me a blank check. They needed to see the product.

So, with a week to prepare, we set up a new call where I would demo the platform. The best thing we came up with was for Napster to create a live mock-up, on top of which Frida would apply a branded layer to keep it "on-brand." Meanwhile, I prepared demo challenges based on the info they had shared with me.

It was merely about re-packaging what they had told me—and because it had all come from them, they loved it.

After successfully navigating some tough questions (such as “Have you done this before?”), we verbally agreed to move forward. In less than a month from my initial outreach, the ink had dried. Bell rung! Even at Cole+Moore, I had not been able to sign this company, and not for lack of trying. Just like that, we had landed one of America’s biggest healthcare companies as our first client.

Although they were the only client on our roster, that didn’t matter—we were invoicing them for tens of thousands of euros. We were not yet as rich as Croesus,[2] but with this deal in the bag, we were extending the life of our company by at least five months. And because we had very few costs, it seemed like a business model we could extend almost indefinitely. After almost two years of banging our heads against the wall, we had earned a massive win and were ready to enjoy our well-deserved breathing room. It was a eureka moment: focusing on large businesses as clients was the way forward.

* * *

Soon after, we went to a tech meet-up that Sting put together as a Friday after-work treat, which was ultimately a gathering of computer geeks and extravagant brats sipping on alcohol-free beers and pretending to be grown-ups. All the cool guys—the who’s who of Sweden’s stellar growth companies—had been invited. Even the tough guys from the most prestigious venture capital firms in town had shown up.[3]

The prime reason everyone had assembled was because Daniel Ek, the founder and CEO of the music streaming service Spotify, was giving a talk about growing a multi-billion-dollar company. We were far from those astronomical numbers—and not just us, but nearly everyone in the room. The companies present were the crème de la crème of the Stockholm startup scene, and yet most had no real income to speak of and very little idea of how to generate any. We'd been like them as well just weeks before when we were courting stars and trying to sell average people a better version of themselves. But now, we were proud that within Sting, we were the outliers.

Aside from the thrill of meeting Daniel Ek, we, the suits, were at the event because we needed Stockholm's startup community to know what ChallengesЯus was pivoting into. I also wanted to get a word with Daniel if I could, but after his very exciting talk and a hurried Q&A session, he zipped out. Clearly, he had more important things to do than hang out with a bunch of wannabes on a Friday night.

While the kids nibbled on carrots sticks dipped in hummus, eagerly discussing the validity of PHP versus Python[4] in the LAMP[5] open-source Web development platform with other techies, Magnus and I were cruising through the room collecting business cards from prospective investors. After receiving our elevator pitch for ChallengesЯus 2.0, most of the companies we spoke to were responding warmly, telling us to add them to our distribution list. We had no such list, but we resolved to put one together first thing Monday. Still, we were about to learn that our new approach came with an avalanche of new problems.

Just like little kids create little problems and big kids generate bigger problems, inking the tenth biggest and most admired company in the world had come with a massive headache. On a pragmatic level, we now had two businesses: a lifestyle and self-improvement business with a funky platform geared toward Gen-Y adrenaline junkies and a rather conservative business aimed at human resources (HR) professionals at the intersection of traditional management consulting and tech.

Our website's landing page clearly did not reflect the latter; we had a carousel of pictures with daredevils, extreme sports addicts and fitness beauties to help us drive traffic. It was not what you wanted to see if you were the HR director of a serious firm and certainly not if you oversaw more than 3,000 healthcare employees. Our entire website needed to change and fast.

Bypassing the extreme athletes and bikini babes on our homepage was our first priority, and once again, Napster delivered. He registered a new domain and linked it to a custom copy of our existing platform. Though it was a good quick fix, we had no history or references within the industry we had just entered, and there were still plenty of traces of our nonsense challenges all over the internet. Our only hope was that no one at our newly acquired client behemoth—not the procurement guys sitting at HQ nor the outsourced IT geniuses working out of Bangalore—would figure out who we were and bring the whole project to a halt.

If our situation was stressful before, now it was outright chaotic. Beyond changing the site, we would have to rethink all our internal processes to align ourselves with our new business strategy. Forget about wellness gurus' diaphragmatic breathing techniques—we were now running around like headless chickens. From the time I got out of bed at around 6 am to when I finally lay back down some 20 hours later, my heart was racing like a Formula One car on the Marina Bay Street Circuit.

9. ANY PRESS

Before we knew it, six months had passed in Sting, and it was time to pitch ourselves again on why we deserved to stay in the incubator. Although we weren't so worried about our chances of surviving a little longer, we were in a new kind of emotional bind. Letting go of our original idea of sparking individual transformations in favor of going back to B2B was painful. Initially, we tried to do it all.

As we explained it in our pitch, in practice, it would mean running two startups in parallel.[1] We tried to convince Peo that we could become a sort of LinkedIn for challenges, with one foot in B2C and another one in B2B. ChallengesЯus.com was my and Magnus's baby, after all, so we wanted to protect it.

"Imagine if we could sell challenges to advertisers through one door and business challenges to corporations through another," I said to Peo. "Would we then, eventually, not be able to leverage cross data and map the complete persona of our users? At work and in their private lives? This could be huge!"

"If there is one thing I have learned in my life as an entrepreneur," Peo said, "it is that you cannot be everything to everyone. You have to pick a horse." Deep down, I knew he was right. I had seen it myself in many successful companies—they had a niche, obvious differentiation and a clear customer profile to go after. In a way, those who we call unicorns were one-trick ponies.

At the end of our pitch, the Sting coaches were all looking at us a little sideways; as ever, we seemed to be all over the place, flailing between business models and trying to do too much, too fast. Still, it was clear we had gotten traction, external funding and a sizable first client. On top of all that, they liked us at Sting. Without much ado, we were accepted into the next phase of the program.

As we started to multiply our meetings with potential investors, we learned the hard way that trying to do different things was not something they liked. Many people simply did not get what we were doing, which we couldn't understand.[2] What we did know is that we needed to generate some positivity around our decision to pivot however we could. Fortunately, the opportunity soon came.

One day, all of Sting's startups got an email. A senior journalist at *Roundup*, a prominent Swedish business journal, was looking for a startup to profile for a big story they were planning. There was a "but": the selected startup would have to agree to fully disclose its progress in the fundraising process. It meant the selected company would need to be incredibly transparent about a process that typically only happened behind closed doors.

To the journalist's knowledge, nobody had ever taken that kind of angle on startup coverage before. It would have to include a real startup looking to raise real cash on one side of the table and a real investment company looking to invest on the other side. Journalists would also be in the room to take pictures and capture the discussion in real time as it unfolded. A handful of companies on Sting's roster met the stringent requirements imposed by *Roundup*. Three applied and only one won: ChallengesЯus.

In all honesty, I don't really know why we won over the others—perhaps it was the nature of our respective backgrounds, namely that Magnus and I were a bit more (to put it delicately) exotic. Both of us were considerably older than the other founders, and we were coming from an industry that wasn't typically known for its out-of-the-box thinking.

Because *Roundup* was a business magazine geared toward executives of medium-to-large companies in Sweden, we also assumed our approach to business had piqued their interest as well. We were a business trying to "cascade strategic initiatives" with a "fun and social" angle in an industry marred by conservatism. It was an angle that fit particularly well with their readership.

What was certain, however, was that we already had a recognizable global name on our client list—an actual paying customer rather than the "pilot projects" startups usually had. It meant the discussions we would have with new investors would likely be meatier and the opportunity more promising for all parties involved.

Although there was a chance that the level of transparency we were agreeing to with *Roundup* could potentially backfire and hurt our credibility with new investors, we figured we had nothing to lose. Our new model looked promising, but we needed more reach and notoriety if we were going to sign new business and scale.

When the interview happened, everything went as planned: we sat down with some investors we'd connected with through Sting and negotiated in front of a panel of *Roundup*'s journalists. They took notes on the proceedings, and ultimately we left the meeting without a clear guarantee that the investors were interested (which was a bit of a lackluster ending to the article, unfortunately). From our perspective, the whole idea to do the interview at all seemed like a bust; the journalists had been pleasant enough, but we couldn't see how the article would help us. Fortunately, we were dead wrong; in fact, we weren't ready for the kind of waves it would create.

* * *

Anders was one of the executive co-founders of Cole+Moore. The two other guys with whom he had started the firm some 30 years ago had long left, but Anders had stayed on. Cole+Moore was a true passion for him. For years, he had always been the first at the office every day. It was one of his habits, and he was very proud of it. He had never missed a day, although he clearly could have. He was a company man, through and through.

A few days after my meeting with Caroline to tell her I was quitting Cole+Moore, I had told Anders about my big plans to become a startup founder. All he had told me was how hard starting a business really was.

"Are you aware that some 137,000 businesses are registered every day?" He asked. "That's 50 million new businesses per year, globally.

"Wow, that's a lot," I replied. "No, I did not know that."

"Can you guess how many will actually make it in the end?" He said with an almost sadistic smile. It seemed like he was trying to impress or intimidate me because he had obviously succeeded with Cole+Moore. I had no clue how to answer, so I just threw out a number.

"I don't know—maybe 20 or 25 percent?" He burst into laughter.

"I know you are very optimistic, my dear friend," he said, "but I am afraid to report that a mere 10 percent of those starting at any given time are still running after ten years.[3] Really, Arnaud, if I were you, I'd stay with Cole+Moore doing what you do best: sales."

His numbers were chilling, but it was not just about money for me. It was about growth. How far could I go on my own? What would I find on the other side? *Was* there even another side? It was about self-exploration, a journey into the unknown. I wanted to defy the odds, and nothing was going to stop me—not the economy, not any scary statistics and certainly not Anders.

Some 20-odd months later, Anders arrived at his desk like he always did at 6:15 am. His daughter was in her 20s and had taken a year off to work as an au pair in Australia. He had no kids to drop off at daycare in the morning; it was just him and his wife now.

He greeted the delivery man, collected the company post and entered the office building after mechanically deactivating the alarm. Then, he headed straight to the coffee corner to a machine he had selected himself long ago. It was a top-of-the-line espresso maker, a DeLonghi Magnifica.[4]

After that, Anders picked up the latest *Roundup* and sat at his desk, ready to start the day. After skimming the magazine, his eyes widened and he rushed out of his corner office, looking for someone to talk to. By 8:30 am, Anders had not calmed down at all.

"What the hell is this?" He asked every other senior executive that morning as they passed his office. "Who on fucking earth let this happen?" His blood was boiling because the *Roundup* article we'd been interviewed for was more than anyone could've expected: it was *eight pages* of coverage with full-size colored photos, many of which depicted Magnus and me grinning as widely as we could.

We had heard from friends and former colleagues at

Cole+Moore that we had been the topic of discussion for weeks. Anders could just not get over the story. He took it personally. Anders himself had never been featured in *Roundup*, despite his loaded personal bank accounts and his expensive tastes. It was why, I presumed, he had been so furious that morning. He just could not stomach it. Though we had clearly not "made it" yet, we were still in a magazine that printed hundreds of thousands of copies.

After reading the story, his takeaway was not that ChallengesЯus was a company of personal change that had recently pivoted to B2B. No, from his perspective, Magnus and I had left to compete directly with him and Cole+Moore, which was a complete betrayal. Of course, he hadn't the slightest idea of the two-year journey we had gone through up to that point. That part, of course, had not made the headlines.

It is often said by old-school capitalists that "nice guys finish last." While our world is a highly competitive place, "survival of the fittest" has never been about being the hardest or the strongest. In fact, winning in nature is about being the most adaptable to the demands of the environment.[5]

Over the years, I had seen people all around me—maybe a majority, even—climb to the top by being blatant douchebags. They exhibited values I would have assumed were against the interests of natural selection, values like integrity, honesty, charity and humility. To reconcile that, I have often concluded that successful people in general may have been nice at some point. I hypothesized that they were hardened over the years, turned into black diamonds by the pressures of heightened responsibility, stress and compounding greed on the road to success.

I took comfort in the idea that being successful at work did not automatically entail being successful in life—and you could be a nice guy and win, too. Either way, we were nice guys, and we had absolutely no intention of finishing last.

10. SHOOTING FROM THE HIP

While we were pleased to hear that our positive press was getting under the skin of some of Cole+Moore's execs, we were less thrilled that the story was making Caroline have some second thoughts.

"The issue now, Arnaud," she said tersely over the phone, "is that I have now financed a startup that has pivoted so drastically that it's now almost directly competing with a company *I still work for*. Do you see how that could be a problem for me?"

I heard her complaints, but what could I say? How could we have anticipated this change? None of it was part of our plan—it had just happened. Pivoting into B2B and selling challenges to companies had been a survival mechanism, and it wasn't something we were entirely thrilled about, either. Magnus and I had started a business to get away from management, change, strategy and PowerPoints, and yet here we were going back to it.

"I understand your perspective," I said, "and I know you'll have to be discreet, but look at this way: your likelihood of getting a great return on investment has substantially improved, if anything. Magnus and I know this area better than *anyone*!" I could practically hear her rolling her eyes on the other end, but she couldn't deny that I had a point.

Even if new investors had been nervous before, it was strange the way things were working themselves out. We'd had our one client and a few pilot projects here and there before the article, but after, we started picking up steam.

Over the course of the year, we were acquiring new clients—among them big names in industries from pulp manufacturing, beverages, consumer products and energy to construction, just to name a few. In all, we had corporate clients using our platform to challenge their employees all over the world—from the US to China. We were not exactly bathing in cash, but we were in the best financial situation we had been in so far.

We were going through our first major transformation as a company, and on some level, it was hard not to feel like we had wasted two years (and perhaps even more in the ideation process that preceded it). But had we really? The truth was we had done what we believed in and thought what we were doing was important. It took us places, and we had loads of fun in the process. We had also learned more about human behavior in practice than we had during our many years at Cole+Moore.

We had tried things, developed theories and drawn our own conclusions. We had tested so many assumptions about driving behavioral change online, validating quite a few of them and discarding so many others. We now knew *precisely* how we would support organizations in their complex "change journeys" through an online medium. Clearly, we would not have been as credible talking about the complexity of working with behavioral change online without our first iteration of challengesЯus.com.[1]

We were moving away from Guinness World Record challenges and our original motivating vision of getting the world's average Joes off their couches. And we were getting away from the ambition of taking down Facebook in the process. But even if we still were keeping one eye on Facebook (maybe we would circle back and take them out later, we figured), the new corporate tech giant in our sights was LinkedIn; they weren't our direct competitor, necessarily, but we were keeping an eye on their moves because they were the only B2B online platform we knew of that could potentially end up in our newly discovered niche.

In a way, being back in B2B felt a little bit comforting. We could finally talk the talk and walk the walk. We had been swimming against the tide with our unconventional business idea for about two years with some level of success, only to go back to a world we thought we had left behind. Although I still nurtured fantasies about being the cool new kid on the block, I was happy to be back in an environment where I felt I had more control—and for once, things seemed to be falling into place. Of course, there was another complication on the way.

As we soon discovered, many of our bigger clients wanted to add a clause to our service agreements preventing us from using their logos in our marketing. I understood why large companies had such clauses, but as a startup, it still felt profoundly unfair—posting work with well-known clients on social media was one of the only ways to build public interest. Sometimes we agreed to those clauses, but often we did not. It all depended on the size of the deal.[2] Our workaround in those cases was to use the tacit knowledge that so many service providers, particularly in the consultancy industry, were so good at leveraging.

The last thing we wanted was a lawsuit from our corporate clients, so when talking to potential buyers, we would simply say that we worked with a Fortune 10 company, a top 100 company or whatever the case may be. Depending on our mood, we sometimes pushed the envelope by saying the companies we worked with had a blue-and-white logo and a yellow-and-red one, for example.

We were tweeting and posting our achievements on LinkedIn as much as we could, but it wasn't getting us anywhere. We were way too small for people to hear us. We needed a much bigger amplifier, and by jumping through all sorts of hoops, we managed to get ourselves another interview with a journalist. After all, the *Roundup* strategy had worked before, so couldn't it work again?

The journalist we spoke to worked for a prominent

online magazine specializing in internal communication. After a quick conversation, our interview was out 10 days later—and once again, Napster had been a genius. He noticed it was possible to keep our story at the top of the online magazine's main page by going there and clicking on it. As a result, that's what our entire team was doing several times a day.

Soon, we turned the whole thing into a game: who could get the most clicks per day? We even had an improvised leader board on the flipchart in our office. It was fun, but after a week or so, we lost interest in the game; instead, we automated the clicking process with a simple program David wrote and stayed on top for weeks.[3] As we'd hoped, the article got us a lot of traction.

Though the direct return on the article—that is, hard cash directly attributable to it—was probably zero, it provided us a fantastic conversation piece with potential customers, investors and employees alike. Still, most of the connections we were making from it were still leading to one-on-one meetings behind closed doors, which I knew wasn't enough to give us the reach we needed.

I was desperately looking for better avenues to get the word out, and getting new clients by leveraging the brands we'd already done work for seemed like the only reasonable way forward. Though we'd said we wouldn't use any of our big clients' branding to boost our own, we intended to stretch people's tacit knowledge to the extreme so that they could guess who we were working with from context clues. *Besides*, we thought, *we aren't being paid enough to stay this silent.*

Grace Murray Hopper, an admiral in the US Navy, once suggested that to be successful in this world, one should ask for forgiveness rather than permission. Admittedly, when a senior officer of one of the most (if not the most) top-down, hierarchical and red-tape-driven organizations in the world gave that advice, it was probably worth listening to. Even if we were still a little attached to our identity as the "nice guys," we decided to take Hopper's advice, pump up our marketing volume and worry about the fallout later.

Because of the many seeds we were planting, a leading podcast in the field of business management approached us. They were keen to air our story, but the case would have to be told from both perspectives: the supplier (us) and the client (whoever we brought to the show). I presented the concept to a few select customers and only those I thought would help position us in the best light possible. Dennis, the HR Director at the monster healthcare company who had first purchased our services, agreed to join us on the show to discuss the process of working with us.[4]

For our gambit of doing a podcast with Dennis to work, we had to be smart; it could not be about challengesЯus.-com, it had to be about Dennis. It was about his organization and his visionary leadership: how they had embraced new technologies despite being a hundred-year-old, 800-pound gorilla.

After we did the show together, it was clear that it was

an absolute triumph. Not only did we get a top-rated company to vouch for us by voicing their genuine satisfaction with our platform on air in front of a global audience, we were also implicitly given carte blanche to market our work going forward. From that point on, the marketing guns were out, and the suits were no longer just going to be nice guys—we were also going to be cowboys shooting from the hip.

Amidst all our marketing efforts, we were in talks with a Fortune 100 company, the US-based version of our giant healthcare client, to use our platform to cascade their new strategy to their 5,000 employees globally. It was a project that was larger than the one that was currently keeping us busy with the European operations of that same organization.

Soon after, we received documentation for a security audit from the company's US headquarters. They wanted an audit to ensure we were worthy of working with them and that we were completely risk-free. I had tried every tactic in the salesman's playbook to avoid the audit; as a four-person shop, I knew we clearly did not have the muscle to handle all their system requirements—at least not on paper.

For reasons I still do not understand, we had skirted an audit in Europe the first time around, and I certainly intended to do the same on this bigger project. I had, for example, offered to provide internal references from the company's European peers. I had also suggested calls rather than a paper audit in the hope of being able to yammer our way out of it, but they were having none of it.

It was clear I had failed when I received an email from the US procurement team with a 40-page questionnaire as an attachment. We were expected to fill it out to the best of our knowledge and ability and send it back with the necessary supporting documents.[5]

The audit paperwork was intimidating, but we still needed to win the deal. If we did, it would somewhat confirm that we could sell to the same company not just once but many times. A repeat purchase was not just a good indicator of the value provided to a customer; it was *the* indicator. It would also signal the scaling potential of our model, for which we were all desperate. There was no way around it: we had to roll up our sleeves and get to work.

The first time I glanced through the audit documents, I couldn't even comprehend most of the questions. Even Magnus, the brainiac between the two of us, was flabbergasted. There was clearly a lot of tech gibberish that only Napster would be able to decipher, but even Napster was reluctant to go through it.

"We can simply not send this back," he said. "There are so many questions I don't even understand. As for those I do understand, we cannot in good faith say we have all the stuff they require." *Was that really the case*, I wondered, *or was Napster simply being his brilliant but irrationally anxious self?* Either way, it was painful to hear. After a long afternoon of deliberation and risk assessment (as well as a night to let it sink in), Magnus and I concluded that we had to try. We were cowboys, not soft toys.

"Let's give it our best shot," Magnus announced to the kids in one of his grand speeches he gave every now and then. "At this stage, we do not have much to lose. Plus, we are not fully tied up anyway, are we?" Although he finished his sentence with a question, we all knew it was 100 percent rhetorical.

Magnus and I took the nontechnical questions, obviously, though there were so many of them that we did not even know where to start. The whole thing felt stupid because the questions simply should not have applied to us. For instance: How many fire extinguishers and alarms are present in the server room? Are all data center workers trained in control and storage of combustible materials? Is all equipment mounted onto labeled racks in accordance with local earthquake guidelines?

We did not have racks, data centers or server rooms. We did not even own a server; we were merely renting one from a supplier via third-party hosting. Presumably if our hosts were to be interviewed, the answers would be yes and yes. The latter was probably a yes also, but there were very few earthquakes in Sweden, and the ones that did occur tended to measure around 3.0 on the Richter scale and typically went unnoticed. The last noticeable one had been more than a hundred years before—in 1904, to be precise. Of course, none of that would be a good enough answer.

Initially, we contacted our hosting partner to get some

clarity on these topics, but for security reasons, they refused to enlighten us. We had no other choice but to make some self-educated guesses while clearly stating that we used third-party hosting. After a good session of creative visualization, we had a good idea of what our partner's data centers must've looked like—how many extinguishers they likely had and so on.[6]

For their part, the kids had to deal with equally thorny problems. They had to do server network topologies,[7] show that all components and data flows in our service had been threat-modeled and sketch our backup and disaster recovery strategy and steps. We also had to discuss our clean desk policy,[8] our certifications, our ISOs[9] and all the other policies huge companies like Microsoft or Oracle would have folders (if not entire rooms) dedicated to. We had none of that—and our desks were far messier than they were clean.

As we went through the documents, we realized a few things. The first was that we had been unbelievably fortunate to get our initial European deal and that perhaps we should have just patted ourselves on the shoulder and savored our limited success rather than pushing our luck further. The second was that these audit questions were clearly meant to get rid of small companies exactly like us.

Like any other Fortune 100 company, our client was a massive machine that could ill afford the risks associated with working with a tiny company like ChallengesЯus. It was probably why startups typically didn't start at the top of the food chain. Instead, they had to sharpen their saws by gaining experience and confidence at the bottom. Slowly but surely, they moved their way up, and by the time they got there (for the few who actually did), business negotiations became a battle of equals with few real logistical issues.

After two full days of focused work, we mailed our answers in. It took a good three weeks before we heard back from them. James, the lead tech procurement officer based in Chicago, Illinois, wanted to have a call with us to ask further questions. Jennifer, our buyer in New Jersey, would also be on the call. *At least they didn't just slam the door on us*, I thought. Despite the half-baked answers we had given, they still wanted to talk. We took that as a good sign.

"The party is *on!*" I howled as I read Jennifer's email inviting us to the next step in the process. It was another mini-win, as we used to call them. We had a massive opportunity within our reach to win yet another mighty company, but before that, we had some difficult decisions to make.

We imagined that the questions they would put forward would, for the most part, be highly technical; after all, we had already won Jennifer's heart with our project management skills, which was why we were having any audit discussions in the first place.

As far as the tech was concerned, we were quite frankly not well suited to answer questions. We knew David could answer tech questions, but we also wondered whether he would screw things up for us even more than we would by being too candid (being the blunt tech genius that he was). For a minute we considered whether we should all be on the call, but that was also delicate.

We had to signal that the call—and the entire deal, for that matter—was not important enough for us to justify using an hour or more of every team member's precious time. In some sense, it was the oldest trick in the book: we had to be enthusiastic, but the last thing we wanted to reveal is that we were needy. Buyers hated needy suppliers. After all, if a supplier was as good as they said they were, why would they be needy?

Rather than being overly accommodating, we had to appear as if we barely had the time to take their call—that they should feel like we were doing them a favor by talking to them because we had more important things to do than spend an evening reassuring them about their purchase. Sales was a psychological game, so it was important to play our cards right.

One thing Magnus and I were bullish on was our ability to speak what I liked to call "artfully vague language." We speculated that in so doing, we would be able to smooth out any corners on the fly and take our deal across the finish line.

The call was to start at 10 pm Central European Time on a Tuesday. Magnus and I were at home, he in Uppsala and I in Stockholm. Our game plan was that if we somehow got stuck on a topic, we would Google it as quickly as possible and hopefully make up something that made sense. We had also agreed to have a side chat open on WhatsApp. We were fired up and ready to rock.

As the call started, the team thanked us for joining so late in the day in Europe and for submitting the audit on time. Things were going smoothly until about 20 minutes into the call until they asked us something about SSL.[10] Napster had taught us about everything there was to know about SSO,[11] but we had never heard of SSL. Even as I googled it and saw the explanation in plain English, I did not know what to do with it.

Napster would have known what to say, but he wasn't there, so we had to think fast. For a moment, I thought I should message David, but there was no time. We should have had the kids on standby via WhatsApp, but it was too late for that. There was a long moment of complete silence that felt like an eternity, during which Magnus and I exchanged side messages. Soon we were brought back to reality in a snap.

"Guys? Are you there?" said Jennifer. Her question lit a light bulb in my head.

"Cou…peat…tha… quest…?" I said.

"Oh Arnaud, the line is bad," Jennifer said. "Could you repeat that please?" I kept up my pantomime that the line was cutting out, not sure of what else to do.[12] "I'm very sorry guys, but we cannot hear you there. Do you hear us?" After more theatrical stammering, we cut the call. There was nothing else to do; nothing smarter came to mind in the moment. A few seconds later, my phone was ringing. It was Magnus.

"I know," I said. "I'm sorry, but that was the only thing I could come up with."

"Are you freaking kidding me? That was pure genius, dude!" Magnus replied. "Trust me, I was about to do the same thing if you hadn't jumped in first!"

At that, we laughed so hard that our abs seized and could hardly stop. After we had finally calmed down, we discussed the call, trying to analyze how it went. Given how it had ended, it was not a straightforward question to answer.

We had to do something to get our negotiations back on track, so we agreed that the most sensible thing was to email them, excuse ourselves for the bad connection and suggest that they send the rest of their questions by email. This way, we would be able to provide them with the answers in writing the following day or so. That was precisely what they did, and it was how we got Napster to nail the last few questions. Though we couldn't believe it ourselves, we successfully passed the audit.

A few days later, we received the 35-page contract. Once again, there were so many clauses we clearly could not live up to, notably the "indemnities" in the millions of dollars if we somehow messed up or lost the company's data. Since this client was involved in many things, from aerospace to medical devices, it was understandable that they had clauses to protect them from business partners losing their valuable data. After all, what if a mistake like that led to the explosion of a space rocket at take-off? Or a medical device not working in an intensive care unit? Still, the penalties seemed steep considering we were selling them software to generate internal ideas and challenges to motivate their employees.

The choice turned out to be really simple: either we took calculated risks, signed the papers and got the deal, which would offer us a chance to prove our model; or we didn't sign the contract and avoided any potential trouble, forgoing our chance to make a dent in the universe.

Ultimately, we did precisely what any startup in the world would do.

* * *

After getting so much good press and landing another whale, ChallengesЯus was on everyone's lips at Sting. We had seemingly become the archetype of a successful go-to-market.

"I've never seen anything like this before," Peo said to us after we told him our latest good news about the audit. "You guys barely have a working product, and yet your topline is already quite decent. I just can't believe you have actual paying customers—and big clients, not just pilot projects!" For our part, we could hardly believe it either.

We still had a handful of clients that were small,

unknown companies,[13] and while they weren't making us any profit, they were helping us market the fact that we were growing, which was, in turn, helping us attract bigger fish. In total we had around six clients, two of which were significant.

Soon enough, the youngsters, hippies, dreamers and techies of Sting were all coming to us for tips on how to break through. More specifically, they were asking us how to price.

Pricing was something vital I had learned from my days of hardcore consultative selling. As I had seen firsthand, unless a client was willing to put real money on the table for a service or product, the product or service may not have real value at all. In our second iteration of ChallengesЯus, we had taken that nugget to heart.

I knew from business school that pricing was a cornerstone of business success. You simply had to get it right. Although it sounds like an aphorism, applying that theory in practice proved rather challenging. Our focus was not on psychological pricing[14] but pricing from another perspective. Pricing largely falls into two categories: cost-plus[15] or value-based.[16] Initially, Magnus and I decided to fully embrace value-based pricing and did some experiments to determine price elasticity.[17] What we discovered was that demand for our service did not necessarily change as we dropped the price; thus, demand was inelastic—but the key takeaway was that it took us *just as much time* to close a tiny deal for almost no money as it did to close a huge one.

The math was clear: obviously, we could only afford to sell high.

With that established, we just had one more thing to figure out: how willing were we to risk losing deals for being too expensive? How far could we stretch our elasticity curve before it snapped back in our faces? The decision was that we would stretch it as far as we possibly could, and the result was that we were making top dollar[18] for our services.

We soon realized we could charge outrageous sums for our services despite having minor costs—it was just about coming up with a number that felt right for the client, articulating value and, when required, making our operation sound overly sophisticated. As we found, we didn't have to worry about complicated Excel formulas to find our price. The way we determined that was simple: raise our price in every pitch we did and notice what the acceptable upper limit was.

Our big corporate clients had more than enough money, so we viewed our pitches as the best product/market fit test money could buy: would major companies buy our services or not? To Peo and at Sting more generally, this seemed like brilliant strategy on our part, but it wasn't entirely. Part of our decision came from pure frustration.

In the face of so many naysayers, we had been continually looking for a proof of concept for our model. It meant that early on, we were saying yes to every customer demand—even if it meant deviating from our developmental roadmap. Before long, we were building custom features for free and telling ourselves we would learn from it, or rationalizing bloating our platform with features that we thought clients down the road might need. It had made sense on paper initially, but it wasn't working.

As we learned, every single customer need was different. None of the special requests we received were the same, so there was no red thread uniting them all. All we were doing, in effect, was devaluing our work. Giving away quality for free felt like a winning growth strategy at first, but as we discovered, it was actually a terrible one. All we had been doing was digging ourselves into a hole that was impossible to climb out of.

The minute we realized this and raised our prices across the board, all our cheap customers who had no intention of paying us fairly for the value we were providing them magically cleared away. In turn, those who genuinely wanted an extra feature or special treatment and understood that we could fix their problems were by and large willing to pay us more.

The result was that our internal development pipe was suddenly unclogged. Frida and David had more time for mission-critical tasks, and Magnus and I could focus more clearly on strategy. The bonus of charging for special requests was that we were essentially using ChallengesЯus's customers to finance the platform's development.

The truth of our success was that we had finally found

the courage to say no and had recognized our value in the process. After that, the seemingly colossal difficulty of choosing what to charge dissolved. It was about self-worth. Were we willing to do all that work for nothing? Were we not worth more than *zero*?

In all, our new philosophy resulted in bigger clients, money coming in and more room to breathe. We were finally on the good end of the deal—for a while.

11. BUSINESS UNFILTERED

We had learned quickly as a young company that the most valuable business events were either events Sting had hand-picked us for or events that organizers chose based on our expertise, niche and maturity. In truth, the latter kinds of events rarely came cheap and were much more of an investment. At first, as we considered networking opportunities and strategies to find more clients and spread our name, we decided that the only events we would consider had to fall in one of those two categories.[1]

Since we had been to business mixers back in our Cole+Moore days, Magnus and I already knew that we had to take the claim that a day of speed meetings with HR directors from large corporations would generate deals with a pinch of salt; at the same time, we also knew that a single deal gained from such an event would dwarf whatever we had to pay to attend. On several occasions, we tried attending those events, distributing leaflets and sharing our elevator pitch with a big smile to any curious soul who approached us. Eventually, we found that the return on our investment was frankly pathetic.

Eventually, we devised a simpler rule that was even more effective: if ChallengesЯus was not to be one of the central companies at a particular event, we would decline the invitation. Like the rule we had developed about custom features for prospective clients, the point was to help us stay focused on our prime objectives and cut through oncoming distractions.

Soon after we had passed our audit and secured another major client, we had been selected with seven other startups to pitch at Sting's own Sting Day, one of the most revered startup events in the Nordics. Close to a thousand people were set to attend.

According to the event organizers, we would have three minutes to present ChallengesЯus to a panel of investors, three from Sweden and two from the UK. After the presentation and questions from the panel, the audience could cast their vote for "pick of the day." The winner would get the honorable award of Rising Tech Star of the Year, a €10,000 prize and all the attention, media coverage and investor preference they could ever desire. After hearing that, our goal was clearly to finish first. The event was in perfect accordance with our rules, but Magnus did not really feel like presenting, so I had to volunteer.

When Sting Day arrived, I marveled at the setup and the enormous stage from where I sat. There were at least six massive projection screens all around the room, two of which were directly mounted high on either side of the stage. On stage were five tall chairs, waiting empty. Background music played until the lights suddenly dimmed and the moderator, Annika, appeared.[2]

After we were warmed up, Pär, Sting's CEO, made his own entrance to a standing ovation. After 20 minutes or so of background, mentioning key figures in Swedish entrepreneurship, explaining Sweden's place in entrepreneurship versus the rest of the world and establishing its role in the "next wave" on the radar (that of the so-called Industry 4.0[3]), Annika took over again. Finally, she called for Aram Sargsyan to join the two of them on stage.

I had never heard his name before, but after seeing his name and company logo flashing on the projector screens, I knew the guy had to be *huge*. With a big smile on his face, Aram rushed through the crowd from the main entrance at the back of the room to Survivor's "Eye of the Tiger" wearing blue jeans, a white T-shirt and a Patagonia trucker cap—a brownish one with a blue shark on the front of it. He made his entrance with all the natural charisma of a Justin Bieber, though I was quite certain he must have rehearsed his entrance to perfection.

Aram was no newbie entrepreneur. He had already built a very successful company in IT security, which he'd sold to Cisco five years prior. On that day, however, as a keynote speaker, he was to take the audience through his latest startup: a digital document-signing SaaS[4] company. His life story was fascinating, particularly considering his humble beginnings coming to Sweden from Armenia.[5]

Aram explained how growing up in a post-Soviet Armenia marred by economic difficulties had made him resilient. He told us how he gotten his first breakthrough and how he could not have done it without his so-called FFFs, which stood for Friends, Family and Fools. As unusual as it sounded, he was entirely serious.

"The FFFs," he explained, "are the very first ones you go to if your goal is to start a company. They are a startup's first source of capital, excluding the money the founders can put on the table themselves. Believe it or not, before your first seed round, there is an 'FFF round,' and without it, most startups today would never have made it."

Mentally, I made a note of Caroline—it was clearly thanks to her that we had gotten where we were. Her trust had taken us a long way, and without her initial capital injection, we may never have gotten started. In truth, no one in their right mind should have given us so much money so early.

After Aram's introduction, we listened to four startups' pitches, Q&A sessions from the panel of investors and comments from the wider world of social media. The room was about as energetic and engaged as it could've been when we were introduced.

"Let us now welcome Arnaud, co-founder of challengesЯus.com," Annika announced in her lively voice.

"Come on guys, let's give him a big hand!" Magnus, who had been quietly sitting beside me, elbowed me.

"This is it," he whispered. "It's our turn!" My heart was thumping violently. As I stood up, I could practically feel the epinephrine and corticotropin[6] leaking from my ears. I should have been used to the stress I was feeling since I had done public speaking before, but this time it felt different, although I was not entirely sure why.

As I walked through a sea of people staring at me, I saw myself in focus on the giant screens as I made my way to the stage. I knew people were clapping because I could see them doing so, but I could not hear anything—the only thing I was aware of, at that point, was Annika and her mic.

Despite some nerves, I did my best to center myself and run through our three-minute pitch, which I had practiced ad nauseam. I had rehearsed for weeks for those three minutes of fame and had found it funny how hard it was to condense everything we did into a few clear, well-delivered sentences. When the pitch was over, the crowd seemed to have understood what we were about, and it was time for investor questions.

Bill Hardeman, an investor from one of the two UK-based companies, was gently smiling as he asked his first question.[7]

"Clearly you are onto something since you are already selling to some substantial companies," Bill said. "But I'm wondering what happens once you've challenged all the employees at a given firm. Don't you think you'll get to a point where people will roll their eyes and throw their arms in the air as they are presented yet another challenge?"

It was a valid point, I had to confess. We had not really thought of it in that light until Bill asked the question. I paused for a minute as I tried to come up with something clever to say.

"That's a very good question Bill, thanks," I replied with full confidence. "I'm not sure about that one, so we'll see when we get there! What I do know is that people are driven by challenges. People love to compete and to reach their full potential—it's woven into the fabric of what it means to be human. On top of that, we've built in so many mechanics to drive people to action that I think it will take a while before people get bored. I think we'll be just fine."

Although I had not really answered his question, he seemed pleased with what he had heard.

"When you say 'mechanics,' do you mean 'game mechanics?'" The question came from Lina Fridhet from Brilliant, a seed venture firm specializing in games and social applications.

"Exactly, Lina," I replied. "We've gamified the whole process to increase engagement, from the moment an employee gets a welcome email to the initiative to how we motivate participation along the way."

"That sounds marvelous. Can you be more specific?" She asked.

"Of course. For example, we've built a virtual medal podium where typical users will always find themselves on the middle step. They see the podium from their profile page, and that visualization motivates them to keep going since people naturally want to move up in position. At the same time, it builds their confidence that they are already doing well. That's just one example."

"Wow, I love that," said another of the investors. "That's clever!"

"Another example you'll find interesting is the real-time countdown we've built in to drive urgency," I continued. "As we know from our past experience in change management, unless there is a 'burning platform,' people will be slow to act."

We exchanged some more questions and answers, but eventually my time was up. I went back to my seat and felt like I had done reasonably well. Magnus also seemed quite pleased.

The rest of the evening flew by. Though we did not finish first, we took comfort in the fact that we were one of only eight promising startups in the whole of Northern Europe to be selected that year. As a bonus, we also exchanged contacts with a lot of investors during the dinner and mingle that followed.

* * *

After completing their entire 18-month curriculum, we left Sting in April 2014. On our way out, I vividly recalled our first meeting with Peo and how exciting it was. Our company had nothing back then, but we had come so far since. We had learned so much, so fast, moving from uninformed optimism to informed pessimism. Before pivoting, we had almost gotten lost in the inhospitable Valley of Despair. Now, we were on the other side. It was not quite nirvana, but it certainly felt much better than the Valley.

It was 2014, and we were officially in our third year of business. We were neither McKinsey nor Cole+Moore, but we were tackling serious and strategically relevant problems in organizations, namely engaging employees in the context of change. We were doing it in a more contemporary way than our *competitors*, and we were aiming at establishing ourselves as cool thought leaders in the field of business transformation.

Let's face it: the consulting industry is filled with both old gurus and junior consultants from Ivy League schools, both applying time-worn thinking and rather ancient tools to modern problems. We wanted to rejuvenate the industry. That feeling and that vision had to come across clearly at every customer touchpoint, from our website and logo to the articles we wrote on LinkedIn. It extended to the tweets we wrote and, naturally, to Magnus and me as figureheads of the company.

To align with our grand idea, we had kept our dress shirts and jackets but broke with convention by wearing them with blue jeans and flashy sneakers. Personally, I alternated between a pair of cyan Kappa streets[8] and a vibrant pink pair of Nike runners. Magnus went with a more classic black or white. When we entered a room full of executives from large corporations we targeted, we achieved the effect we wanted: polarization.

The idea was simple: as fantastic as it may sound to be liked by everyone all the time, it was an unrealistic proposition. In general, avoiding external disapproval or negative attention tended to make people completely vanilla. In doing so, they started to blend into the background.

The sad truth was that most average folks might be decently liked by many people, but most do not have raving fans. Not having a fan club is okay if you are alone on a deserted island, for example, but that tended not to be the case in real life. Given that companies were mere collections of people, blending in didn't work in a corporate context, either.

Companies needed fan clubs, and awesome companies had hordes of devotees they nurtured. So, the question was: Did we want to be the beige startup that had no haters but no devotees? Or would we rather break a few eggs, polarize and have cheerleaders and non-fans alike? We chose to be memorable. It was the only way to break through. After all, nobody cared about the 50th most successful consulting firm in the world—people might care, however, about the top three.

In his book *The Purple Cow*, American author and former dot-com business executive Seth Godin talked about *otaku*, a Japanese word meaning "a young person obsessed with some particular aspect of popular culture." The literal

Japanese translation for *otaku* is "your house," as in the place in your mind where your obsessive interests reside.

We all have *otakus*. Some people like Starbucks coffee; others prefer collecting scolopendria.[9] When we first launched challengesЯus.com, our *otaku* was "people who want to challenge themselves and their friends for fun or self-improvement." Through our own experiments, we knew that people joined movements they were passionate about, and in doing so, they connected with like-minded people.

People liked other people who were interesting, unique and who polarized them with their ideas, skills and crafts. It was why we went all-in when it came to our personal brand and positioning in the dry world of business consulting—no matter how that world reacted. And slowly but surely, our strategy worked.

Articles in the papers led to radio appearances and invitations to speak at events, and at each one, we showed up to stand out. Soon, we were *everywhere*. I was being flown across Europe to deliver our story and was becoming the talking head of ChallengesЯus in my own right.[10]

It turned out we had a story that many were keen to hear, and the more I told it, the more people wanted to hear it. We had so many angles. There was the obvious story of an unorthodox young company disrupting corporate consulting with a new social application that touched thousands of lives; there was the behavioral story of how individual people were needed to create organizational change;

and, of course, there was the standard story of the ups and downs of the startup process. The more I went at it, the more we got in return to hype our company. It was a self-reinforcing loop.

After a handful of media junkets, I was contacted by the Learning Department at the University of Leicester. They had been looking in their alumni network for former students who exhibited "apparent proof of success," as they put it.

"We love your story and thought it might be quite valuable for business students considering what to do with their lives," one of their representatives told me over the phone. "We would love to have you come to speak to them in person." It was how I set foot onto the actual campus of my alma mater in Summer 2014 for the first time, seven years after I graduated.

I had gone from being an average, intellectually insecure kid in the suburbs of Paris to getting a master's degree in business administration from a satellite campus in Thailand (all with pretty poor English skills at the time) to this. It was an odd feeling to say the least, but I was so proud.

I was to deliver a two-hour lecture to students in the last year of their MBA programs; that was pretty much all I knew of what was to come. I had titled my talk "Business Unfiltered," where "unfiltered" was a more academically correct way of saying "no bullshit." I wanted my delivery to be as genuine as possible—exactly what I would have wanted to listen to if I had been sitting there myself.

I wanted my audience to get the behind-the-scenes version of starting a company, not the technicalities of it or the traits, skills and mindset needed. No, I wanted to talk about the things nobody ever heard or read about; about failing, being depressed, about being forced to embrace failure and pivot—all the things I wished someone had told me before I embarked on the journey myself.

For days, I had been wondering how to start my presentation. I knew from experience[11] that the first two minutes were crucial for captivating the audience, so I wanted a powerful entrance. Should I storm the stage like Steve Balmer famously did when as CEO of Microsoft?[12] I figured it wouldn't work for me. What would I yell, "Students, Students, Students"? I needed something ending in "-ers" if people were to understand the reference, but what? It would be more in line with the diehard, no-guts, no-glory entrepreneur I was impersonating to yell, "Motherfuckers, motherfuckers, motherfuckers!" as I entered, but as I pictured myself being escorted out of the auditorium by security and my diploma being rescinded, I decided against it.

Could I pull off an Aram-style entrance with music and lights instead? That would certainly be cool, but how? To ask the person in charge of the venue to make such an arrangement would probably make me look like a complete nutjob. As if the universe could hear my internal debate, it served me with the entrance I needed (or perhaps the one I deserved).

It started when I almost missed my flight from Arlanda, Stockholm to Heathrow, London. I had become used to traveling within Europe, and consequently, I showed up at Arlanda airport with only my driver's license, which is enough in the EU. I had overlooked the fact that the UK was outside the Schengen Area,[13] and so the rules were different. It meant that I needed a passport.

Next thing I knew, I was running like a mad man across the airport. I had 25 minutes to grab a cab, explain my situation to the airport police, get a temporary passport issued in express and bolt back before boarding closed. Though I was the last passenger onboard, totally out of breath and sweating like a pig for a good hour or so into the flight (much to the chagrin of the fellow sitting beside me), I made it. If I had missed my flight, I would have almost certainly been too late to speak to the MBA students, as I still had a rather long train trip to Leicester ahead of me upon landing.

The University of Leicester had promoted my lecture for weeks ahead of my arrival, and despite being in the middle of August with record high temperatures in the UK, over 100 students filled the air-conditioned auditorium. Judging by the looks of some of the older audience members, there seemed to be quite a few professors present as well.

When it was nearly time for the event to begin, a voice came through the massive auditorium speakers announcing the arrival of Dr. Matthew Higgins, director of flexible learning (who was, in fact, the person who had arranged the entire thing) and Dr. Noemi McCulley, the dean of the school of business. They sat down in the first row, which had been kept empty for them.

It was only as they walked past the lectern and gave me a solemn look that I felt the pressure of the situation. Right there, I freaked out. Sure, I had presented many times before to audiences of executives at all levels, and sure, I had done Sting Day. But this was different.

The stage on which I stood was regularly host to academics with deep theoretical knowledge. Today, I was there to flash some stupid pictures of apes and of a drenched George Clooney braving a gigantic wave coming straight at him from the film *The Perfect Storm.*[14] *Maybe some of the stories and swear words I was going to say are not such a brilliant idea after all*, I thought. The rest of the audience were not stupid, either; I could remember when I was like them, trying to deconstruct every established theory just for the sake of exercising my brain. Would they ask me to elaborate on Clayton Christensen's innovator's dilemma[15] in relation to what we were trying to achieve? Would they have thorny questions about discounted cash flow?[16] I was hoping not, as I was no CFO by any stretch.

Questions could come from any angle, which was unnerving. Of course, there was no way I could have prepared myself for every curveball they could throw at me—but I still hoped to avoid utter ridicule. Maybe I could pretend to faint? Or leave immediately after it was over ("doing a runner" as they said in the UK)?

I tried to remind myself that these were just dark thoughts bubbling up from my amygdala. It was all just a physiological reaction to a perceived threat. Although there was no saber-toothed tiger coming after me, the hyper-arousal was real. Soon enough, I was called onstage, and it was time for my talk.

"To begin, I want to talk about having a Minimum Viable Product,"[17] I said. "It is probably one of the most used acronyms in the tech world. An MVP is the version of a new product a team uses to collect the maximum amount of validated learning about customers with the least possible effort." After that, I spoke (at length) about how we had failed to develop one in our earliest stages.

"Also important to understand is the difference between traction and scalability," I continued. "Traction and scalability are terms that can seem obscure when you're new to the world of entrepreneurship—they were to me, at least. While they are different concepts, they are interrelated and interdependent. To be successful with a venture, one cannot have one and not the other."

As I explained, an easy way to describe the problem was that you got traction with a good product or service, but you got scalability from a good business model. Unlike the chicken and the egg argument, however, it was clear which one had to come first.

"ChallengesЯus, for example," I said, "got plenty of

traction initially when it launched, but then it plateaued when we couldn't trigger people to complete all the challenges they started. Thus, it failed to spread as we'd hoped it would." I went on to say that after pivoting, we had found a new way to spread and scale successfully, which was not entirely true. Though I didn't say it onstage, the truth was that even as I was onstage, we were still learning the difference between traction and scalability the hard way.

In our latest iteration with about half a dozen substantial projects running, we were already at capacity. Magnus was stuck managing projects, the kids were busy with Boolean logic[18] and I was running down clients to the point of exhaustion, day-in and day-out. Objectively speaking, there was not a lot of room to grow much further.

In every discussion I was having with prospective customers, I was starting to dread what would happen if I got them to sign. Of course, we would accept their business because we were so eager to grow, but I knew it would not be pretty. To keep up, our only option would have been to "sleep faster" as Arnold Schwarzenegger famously put it. In short, we were not scalable—no matter what I was saying on stage.

It was not that our platform wasn't doing its job; it really was. The problem was that we were too involved with every big new client we signed. On top of that, we were still giving customers too many options, despite our mutual agreement to cut back on doing so, and it was killing us.[19] Compounding our problems was that I was the de-facto head of the marketing department, which was at least part of the reason why I was onstage at that very moment.

Even at our current level of success, we were fighting to the death for every new inch of ground. It felt like we were the poilus of the First World War, the French rank-and-file soldiers in the mud with their heads down, marching into a gunfight equipped only with bayonets. Of course, there were limits to how much I could crush the students' spirits in the two hours I had to speak to them. Telling them everything would've been indecent.

After my presentation, which had gone quite well, there was a 35-minute Q&A that also went off without a hitch. There weren't any theoretical business or accounting questions that revealed me as a fraud, and nobody ran onstage to pull my pants down at the last second.

As I packed up my gear, a few students even came to me for selfies. It was completely unexpected and made me feel like a rock star. *Maybe this is what it's like to be Aram Sargsyan*, I thought to myself. To this day, I still remember a comment I got from one of the students: "Cool shoes you have, by the way!"

12. MALEDICTION

"I think we should consider taking external investments," I said to Magnus one day. I was trying to remain optimistic that we were gaining traction progress, but Magnus was not convinced.

Caroline, on the other hand, still thought we were in decent shape. In her eyes, we were truly starting to resemble a real company. Even if we had many small clients who were eating up our time and money, we were also acquiring big clients quite quickly—certainly faster than I ever remembered doing at Cole+Moore. And although Caroline recognized that our fees weren't all that high compared to the rest of our industry, it was also clear that our costs were so low that the road to long-term profitability seemed nearly inevitable.[1]

"Getting nine, 12 or even 18 deals is not an exponential trend," Magnus continued. "These are mere data points. We need more before we can justify raising big bucks for our growth—way more." He was right, of course, but how could we do more than what we were doing without marketing money? Fortunately, Caroline had a good friend who was the chief financial officer at a boutique investment firm in Stockholm.

"The firm specializes in commercial real estate," she said, "but they are willing to invest in good opportunities when they see them, even outside their core. They might be interested." As she explained, the company had previously invested in a book publishing business and a project management app. Despite Caroline's abiding optimism, the hard truth was we needed money however we could get it, so we had nothing to lose by meeting with them, pitching and seeing where it went.

On the day of the meeting, we were greeted by a short middle-aged man with a broad and seemingly genuine smile on his face. To balance his obvious lack of hair, he had a stubbly gray beard, though in it, I could still spot some orange patches. He extended his hand.

"I am Jonas," he said in his tenor, delicately shaking my hand and then Magnus's. "Caroline told me a great deal about you two." For a finance guy, Jonas looked quite friendly. He was wearing a three-piece suit made of navy-blue checkered fabric and a silver tie. It was an uncommon outfit for Sweden, which is still known for its rather casual office dress code. I wasn't sure about Magnus, but I suddenly felt awkwardly underdressed.

As we walked through the modern industrial design in

the firm's office, we saw three other men in their mid-50s seated around a metallic, glass table. The office was in the dead center of Stockholm and had once been a four-story residential building. The floor and attic of the old building had been converted into the sleek loft we were walking through, seemingly a perfect environment for prospective clients looking to buy premium commercial real estate.

Magnus and I shook hands with the other men at the table and sat down. Unlike Jonas, all of them had strong handshakes. The boss at the table, Klas, had deep creases running across his square face. He had the worn-out look of a guy who was not easily impressed, and as we sat, I made a mental note to approach him differently than the other three men—especially if he was the one calling the shots.

In fitting with the cowboy ethos we were trying to maintain, we had not prepared much ahead of the meeting—just enough to have our story straight. The last words I'd said to Magnus as we walked up the stairs were, "Let's be uber-confident." By any external measure of success, we had achieved quite a bit in a few years, so prepping for meetings was not high on our list of priorities. We had confidence to spare, but not time.

In meetings together, Magnus and I typically played good cop, bad cop. It was a tactic anyone who watches police thrillers knows about: two cops question subjects from opposite psychological temperatures—one friendly, one not—with the purpose of extracting as much information as possible. In our case, we were trying to generate sales.

For this meeting, our plan was for me to paint an overly positive picture of what the future would look like with ChallengesЯus on the market. Magnus would then bring my euphoric vision down to earth with pragmatic action plans backed by watertight financial projections.

Our aim was to offer a well-balanced account of our company's past, present and future and to demonstrate that we were not only aligned with each other but that we also had complementary skill sets. As all savvy investors knew, the team was the most crucial part of the entrepreneurial puzzle. Even the best idea in the world could fail without the right team to execute it. Our dreamer vs. pragmatist approach had worked well for us in the past, so using it in our first real investment meeting outside the Sting ecosystem felt like a smart move.

After 30 or 40 minutes of high-energy babble, we seemed to be winning Jonas and his partners' hearts—but I was still undecided when it came to Klas. He had posed some pointed questions that we had answered well, but it was clear he wasn't fully onboard. The other three were essentially gobbling up every word that came out of our mouths.

At the end of the meeting, the men gave us no clues whatsoever about their investment plans.

"We have all we need to make an informed decision,"

Klas said. After walking us back down to the atrium, Jonas told us we would hear back from them soon.

We had a call with Caroline a few days after our meeting, and she had told us that they "liked us more than they liked what we did," which we were unsure how to take. It only took a week or so before the phone rang, which was, quite frankly, surprising. We had been told that these kinds of decisions could take time, sometimes even up to three months. *It must be a no*, I thought as I picked up the phone. *Why else would they respond so quickly?*

After a few words about the weather, Jonas solemnly announced the firm's decision.

"We have decided to invest in you," Jonas said. I couldn't believe my ears! Though we had gone in confident, we certainly hadn't expected an answer in a week. As we soon realized, we had not really accounted for what we would do next if we *were* to get an investment. Our expenses and costs were established and logged in an Excel file, but it was more an issue of us as a team. How would we celebrate this massive milestone? Where would we go next?

Within a couple of weeks, we received our long-awaited term sheet—the most sought-after document for any entrepreneur. After about 10 days of due diligence, we had successfully completed our first "early stage," known as a "seed round." Now, we were poised to take over the world under an exciting new banner: we would democratize change management. More simply, we were going to make change management cool again.

* * *

The entire time I worked on ChallengesЯus, Claire was still working for Sida—and the most important thing about Sida was that employees had to have their hearts in their work. Claire had been pushing papers at Sida's headquarters for way too long, if you asked her, and she was tired of it. One evening, however, she came back from work with big news.

"Sida is offering me a new job!" she said as brightly as can be. It was showing the same sort of excitement I'd felt when we got accepted to Sting or when we signed our first real paying customer, which was great to see. There was just one slight problem: we would have to relocate, and not to another part of Sweden. No, we would have to move to Africa—Mozambique, to be precise.

Moving to Africa would be a huge step—a gigantic step, really, both as a family and for ChallengesЯus. How would we drive the business with me down there? How would I conduct my sales? Would it even be possible? What would Magnus say? What about Caroline? And what about Jonas and Klas, for that matter? I was not certain they would like the idea of me traveling 10,000 kilometers away right after their money had hit our bank account. We lived in an age of connectivity, but there were still limits.

After looking up pictures online, what I saw certainly looked amazing. There was great weather all year round, fantastic deserted white sand beaches, great fish and seafood, a vibrant local culture and so forth. Still, to live there? With kids and an 18-hour flight away from home? I was not too sure.

Beyond the nice pictures on Google was the other reality that Claire did not work for a luxurious hotel group. Mozambique was at the bottom of the UNDP human development index. The people there were dirt poor, and the country had a sad history of conflict and bloodshed. There was near-permanent political instability, periods of social unrest, organized crime networks taking advantage of the country's weak systems, kidnappings for ransom and a very poor human rights record with limited civil liberties such as freedom of expression.

The list of challenges went on and on. It was one of the most HIV-ravaged countries in the world. There were frequent cyclones there. The healthcare system was poor. It was home to some of the deadliest snakes in the world, including the Black Mamba and the Mozambican Spitting Cobra. Of course, some version of all the above was why Sida wanted Claire there in the first place—it wasn't for a sunny vacation.

After thinking it all over, I was not too keen on the idea. Even so, Claire was just like me: she had a drive and a passion for what she did and wanted to make a difference.

Gradually, she convinced me it would be a great opportunity for our family.

"We could give it a try and see what happens?" she suggested. Her "once in a lifetime family experience" angle made for a good sales pitch, but "giving it a try" had complicated real-world implications. After weeks of debating pros and cons, we agreed to give it a try. I would commute, which meant filling my calendar with work-related appointments from Mozambique and traveling back to Stockholm as often as needed to attend them.

Getting ready for the big move was surreal, and it took a good three months to finally get ready. Eventually, we were all packed up, and our things were shipped a few weeks prior to our move. I was not sure if it was a sign from heaven (only the future would tell), but the very last thing I did before closing the door to our suburban house was to ink a new customer. I had done it while sitting on the same kitchen bar stool I'd sat on when our website had gone live.

I was finalizing the client agreement as our car arrived. It was the mother of all taxis: a big, black seven-seater minibus (which, with two pieces of luggage each plus hand luggage, we clearly needed). I sent the contract with our general user terms and conditions document, closed my laptop and we were off.

Although we talked about the move a lot, I was not quite sure that our children really understood what it all meant—at least not Melvin, who had no clue what was happening. Even so, the kids were still happy and excited.

"Africa, here we come, baby!" Melvin was yelling from the back of the taxi.[2] As we gained speed on the motorway on the way to the airport, Claire and I sat quietly looking out the window. We were as anxious as we were cautiously excited thinking about this new chapter in our lives. There was a lot on the line—but wasn't taking risks what made life worth living?

* * *

Setting up ChallengesЯus's African operations during Summer 2015 turned out to be much easier than I anticipated. There was not much difference between me phone-prospecting from Maputo, the capital of Mozambique, and me doing the same from my old office in Kista, Sweden. Thanks to Skype, it was not even more expensive. There was still the same two or more months of delay between establishing contact with prospective buyers and building trust over a couple of calls and agreeing to meet to further discuss the possibility of doing something together.

Sida provided us with a three-level, colonial Portuguese house,[3] and I established my office on the top floor. The house must've been built in the 1950s and had some unique features, including colorful, flowery Portuguese tiles on the walls of the kitchen and bathrooms. There were also some spectacular moldings in every room, as well as an amazing wooden parquet floor made from beautiful tropical hardwood. But all was not rosy.

We had black mold in many of the rooms and countless leaks. During the rainy season, I ran around the house with buckets to prevent us being full to the rafters. We also had all sorts of critters living with us, from regular African behemoth cockroaches to giant wasps who made occasional intrusions on the kitchen. There were also disturbingly resilient armies of ants marching to and from dry storage, where we always kept enough food to survive in case of food shortages, street riots or—as has happened from time to time—a military coup.

The floor was all buckled from old water damage, and things tended to break down every other day. The air-conditioning, for example, had a life of its own; sometimes it worked, often it did not. On any given day, our house was full of more plumbers, electricians and fixers than we had employees at ChallengesЯus. Things that we had taken for granted in the West, like constant access to electricity or water, did not yet exist even in the capital Maputo. So, we had a massive generator to help us out during power cuts and an underground garden reservoir, from which water could be pumped to compensate during hours when water was not available to the rest of the city's inhabitants.

This was certainly a luxury that only the wealthy in Mozambican society could enjoy, and it helped our life feel somewhat more normal. I looked at the generator. It was as old as I was. It would easily have filled a huge room if it had been inside the house. Luckily, it stood at the back of the rather small garden. As could be expected, this ABB-manufactured[4] rusty box released plumes of toxic smoke every time it kicked on, which it did on average five times a day.

Every time there was a power outage, my internet connection would go down. If I happened to be on a call at that moment, the call would drop. I would typically call the person back, and we would continue our discussion as if nothing had happened. There was only one issue in those cases: the generator made so much noise that I could barely hear the person on the other end, even with all the windows closed.

When it did not sound like I was in a busy train station, which was the case when the generator was on, I had other challenges with sounds. At times, it must have sounded like I was in the tropical section of a zoo. Visiting me regularly were all sorts of noisy birds. From my desk on the third floor, I had direct access to the terrace where I could see them engaging in complex dances that seemed like mating rituals.

Whenever I was on the phone, I tried to mute myself as often as possible to avoid any exotic jungle-esque distractions. Maybe it was all in my head, but I wanted to avoid at all costs any exchange that would reveal where I was physically located most of each month. If I were in a leadership position at a large corporation and a random guy called me from Africa to pitch me some sort of web-based activities to challenge my staff, I would probably think it was a joke at best. At worst, I would've thought it was a scam like those typical Nigerian fraudster emails everyone has received at some point:

Dear Sir, SEEKING YOUR IMMEDIATE ASSISTANCE. Please permit me to make your acquaintance in so informal a manner. This is necessitated by my urgent need to reach a dependable and trustworthy foreign partner. This request may seem strange and unsolicited, but I crave your indulgence and pray that you view it seriously. My name is Dr. Stephan Zakwako of the small Kingdom of M'Bobomgia and one of the close aides to the former king of M'Bobomgia...

...and so on and so forth.

Ahead of any video calls, I changed from my casual worn-out T-shirt to a nicely ironed long-sleeve shirt, though I kept my shorts on. No one could see below my waist, anyway. Mozambique was so hot and humid that from the moment I woke up, I was drenched. It was not the sort of look most European callers expected—not even in the middle of the summer, and certainly not in Sweden. Whenever possible, I would blast some cold air ahead of my calls to bring the temperature to an acceptable level.

The light was another thing I had to adjust to over time, particularly from October through February—and especially if I was speaking with folks in the Nordic countries where it was pitch-black most of the day at this time of the year. Through trial and error, I developed a very effective hack with improvised thick curtains that I would hang in front of the windows that were facing the terrace. I think I did quite well because no one ever said anything that suggested they knew I was not in Sweden.

* * *

In 2016, our second year in Mozambique, we started to experiment with alternative sales channels to accelerate our market penetration. In a relatively short period, we had agreements in place with many small businesses around the globe in fields ranging from IT system integration and management to HR and internal communications. Our partners in the UK, the Czech Republic, Austria, Singapore and South Africa all had a ChallengesЯus badge on their site, a testimony to their association with us.

I personally trained each partner on the product and all the whys and the hows of ChallengesЯus. It took a lot of time, but the tests were worth running. The idea was to assess whether someone other than me and our two sales reps could sell our challenges to corporations. Our partners would sell for us and obtain commissions on those sales; if those channels were proven effective, we would certainly be able to scale more quickly and simultaneously across all continents.

At that point, we were lacking partner representation in

North America. In the spirit of not putting all our eggs in one basket, I was having a dialogue with potential partners of all sizes and trades. In each pitch, the angle, positioning, pricing and retainers were all different. Even the depth of our collaborations was different because they varied by the size, geographical presence, impact and potential of the partner, as well as the partner's willingness to work with us.

After over a year of talks with various roles—such as department heads, VPs and managing partners on all continents—within one particular organization in the Big Four accounting firms, we had finally been granted airtime with the US organization's leadership team. Even the CEO, Mike, was on the meeting's participants list.

The call was scheduled during their afternoon, which was midnight for me and late evening for Magnus. We were to demonstrate our capabilities by doing a virtual demo, which in practice meant I was to share my computer screen with all of them. It was beyond exciting, and I had planned every detail of the call very carefully. We had even signed a non-disclosure agreement as they even seemed to have an appetite for a broader partnership.[5]

"This could be the turning point for us," I told Magnus before the call. "This could well be the moment we have all been working toward for years now."

At ChallengesЯus, we had been bootstrapping the shit out of our business from day one, and it was something we were extremely proud of. It's not that we were stingy or too cautious; it was rather the opposite. We understood the concept of cash flow very well, appreciated that we had tax obligations and above all, we took full responsibility for the people we employed.

We also took inspiration from people who had succeeded precisely by taking a no-frills approach to the management of their operations. We often quoted people like the late Sam Walton, best known for founding Walmart and Sam's Club, or the Swedish business magnate Ingvar Kamprad.[6]

Systematically, we looked for efficiency and cost-optimization opportunities everywhere we could in our operations. I must admit that sometimes it was borderline annoying, but I trusted my CFO, Magnus. I must also confess that I would rather have emulated Branson's cash-addled eccentricity—flying hot air balloons and jumping out of planes—rather than Walton's prudence. Nevertheless, although we had now raised quite a sum of money and were making reasonable topline revenues, we still had not invested in an anti-spam email solution, for example.

When we were tiny and had no cash in our coffers, we had calculated that it made more financial sense for us to sort our own emails than to invest in an overpriced anti-spam solution. Since those days, however, we had grown from two to six people and were now getting daily torrents of junk mail. Our tempo had also accelerated exponentially. If we could validate our go-to-market strategy, we could start recruiting more sales resources. Only then, we had concluded, would we review all our internal practices. Until then, we didn't have time.

Life was simply too hectic. I usually fired up my laptop at 5:30 am and worked a solid hour before the rest of the family emerged from their bedrooms; at night, I did not normally hit the sack before 1:30 am. It was not a very healthy lifestyle. *I'll have time to sleep when I'm dead*, I thought. *And before that, when I get rich, I will have the time to get in the best shape of my life.* It was my basic model of the world at that time. The only things that were non-negotiable in my daily routine were: my one hour of physical activity—I either ran or hit the gym—and a few hours of complete devotion to my kids and wife before and after dinner. But once everyone was in bed, I went back to my laptop.

At the start of our call, everything was going according to the plan, and we were having a very good exchange of ideas. They were four on their side; on ours, it was just Magnus and me. As we were demonstrating some specific features of our tool, while babbling about "challenges" as a mechanism to engage employees in organizational change processes, something unplanned happened: an email popped up at the top right corner of my screen—the same screen I was sharing with everyone else.

I saw it as soon as it entered my field of vision even though I was super focused, so I couldn't imagine the others hadn't seen it; after all, they were *only* focusing on the screen, not also on presenting. It read something like "XXX Hardcore chats near you now." Then, it vanished.

I tried to pretend nothing had happened and continued with my presentation. I knew it was a risky move, but I felt it was the best thing to do. I had seen the same diversion in *Father Ted*, the notorious sitcom from the late '90s featuring three priests living on the remote island of Craggy in Ireland. In one episode, Father Ted, the main character, had lost a bet and had to kick his bishop, Brennan, up the ass.[7]

As I tried to stay focused on what was coming out of my mouth, half of my brain could not stop thinking about what Mike's next move would be. Initially, there was no move; instead, a very uncomfortable silence filled the virtual room. It did not take long before one of Mike's VPs excused herself, saying she had another call to attend to. Mike and his peers stayed on, but as we reached the end of our demo, they didn't have many questions for us. We agreed to follow up with a recap and a few possible time slots for a future discussion.

Sure enough, after months of sending emails into a black hole, I received an email one day that our project was no longer a priority and that they would return "when the timing was more appropriate." Although I had heard that same line innumerable times before, I still wanted to believe that they would return to us at some point. I had faith. Of course, they never did.

It bothered me. How could we have gone from high

interest to nothing so quickly? It had to have been the XXX spam email. Our demo with the top dogs, which was supposed to be a final formality before testing a potential collaboration, had been the cause. Just like Brennan in *Father Ted*, we had gotten the boot. Of course, immediately afterward, we invested in top-of-the-line anti-spam software.

I kept asking myself what I would've done differently if I could go back and avoid that disastrous episode. Ultimately, I decided that I wouldn't have changed anything. It had been shameful, and yes, we had wasted a massive opportunity, but we also learned from the experience. The main takeaway was that we could have said something like, "Oops, sorry for that—we run without an anti-spam," even if that would not have come across too well for a company that works with software.

On a higher strategic level, we hoped to use the mishap to our advantage in the future. Ultimately, our error would become a story that well illustrated the values we held dear: be a maverick and have fun. Even so, we couldn't believe that every time we almost took the next step forward, we seemed to get knocked backward twice as hard. It had become so normal that we stopped questioning it.[8]

* * *

If I had not gone into business, I would have studied neuroscience because occasionally I wondered if I was wired properly. In the space of a day, particularly as things got harder, my mind could vacillate between the absolute certainty that ChallengesЯus would take the world by storm and a deep inner skepticism of our (or at least my) ability to convince even one more potential business partner to join us. I could swing from total euphoria to shocking lethargy and depression. It was the startup journey writ large, and it was a roller coaster every single day.

On good days, I swore by Max Weber's ideology.[9] On the next, I was more about Horace's "carpe diem."[10] I could not go as far as Weber that hard work, thrift and efficiency in themselves had the potential to take a person to eternal salvation, but I could certainly conceive that the absence of those values, at least in a company context, would fast track a business to hell.

If Weber has been around at the time, he would probably have had a captivating YouTube channel that I would've followed closely. Instead, I was turning to motivational videos that artfully weaved together scenes of overly buff guys (typically in the mud) with dramatic soundtracks and male voiceovers saying things like "you need to want it so bad that you can't even breathe."

I didn't listen to Adam Smith,[11] the Scottish economist; I listened to Will Smith, the American actor, producer and rapper, explaining how he attributed his success to his willingness to outwork everyone else.[12] I embraced David Goggins's 40 percent rule. The rule, according to the American ultrarunner and former Navy SEAL, says that you are barely halfway to your maximum effort when your brain starts shrieking at you to give up. I don't know the exact percentage, but I certainly concurred that we tended to sell ourselves short of what we could achieve.

I recalled Muhammed Ali's famous line "I'm gonna show you how great I am" and wondered which "you" he was referring to. Was "you" the public? Was "you" his next opponent? I suspect he might have simply been talking to himself—that "you" was his own brain. Ali knew that his brain, like anyone's, would run him over if he didn't pay close attention to it. I guess he knew he had to knock out his own mind before he could even think about knocking out someone else.

Unlike the brain, I kept telling myself that our bodies fed on discomfort and thrived on challenges. It knew no limits. In prehistoric times, we used to run for days at a time on an empty stomach to find food.[13] It made me remember my early exchanges with Eric Larsen, the Antarctica explorer we had first signed up to ChallengesЯus.

Larsen had insisted that he was "just a normal guy trying his best." Undoubtedly, he could lower his heart rate to preserve his energy, but he hadn't been born with that gift. Larsen was no bear going into hibernation. It was unlikely that he started with a higher tolerance to pain than anyone else. More likely, Larsen had learned those skills through years of dedicated practice. Larsen was no more superhuman than me or anyone else; he just had a stronger will than most people and trained daily to stay in control of his brain. Somewhere between absorbing daily inspirational content, being worn down by ChallengesЯus, knowing I needed to become stronger and wanting to escape the chaos of my life at the time, I decided to sign up for a marathon.

I had never even done a half marathon before, let alone a full one. I was not a runner. If I had pushed myself to run eight kilometers on a sunny Sunday, I considered it quite a feat.

It was to take place in South Africa, and I had seven weeks to get ready for it—to prepare my body and mind for an insane 42.2-kilometer race instead of my regular five-kilometer jogs around the block. As soon as I started training, I came face to face with my own squirrelly mind. I began to see myself—the real Arnaud—and my brain as separate entities.

My brain would play tricks on me, coming up with all sorts of reasons to keep me from challenging my habits —*its* habits. Before we even got started, it would say, "You know that you need at least 12 weeks to train for a marathon, right? How are you going to make it in seven?" Other days, it would say, "Why run today when you are already tired? Let's just skip today's practice and enjoy a glass of wine instead," or, "It's way too hot today for a long run."

Even when I told my brain to shut up and convinced it to go running with me anyway, it would still try to argue with me to keep me from doing anything new (and as a result, harder than it was used to). During any typical run, no matter the distance or intensity, my brain would signal pain in my foot, my knee or my shoulder (*The shoulder? Seriously?* I would think). Or it would try to convince me that at some certain pace, I would *definitely* not have enough water in my CamelBak to run a full 19 km.

Ultimately, I couldn't blame my brain. Laziness and pain avoidance were wonderful evolutionary mechanisms. The brain had served humanity well in the past because sticking to well-proven habits often meant the difference between life and death. Nowadays, as I was seeing more and more clearly, there were fewer real advantages to being overly cautious. There usually weren't ferocious animals waiting to eat you between the shampoo and the milk aisles.

In a much safer context than our bodies were historically used to, our brains were actually impeding us from growing by trying to protect us at all costs. They were stopping us from becoming greater versions of ourselves. It was the very same psychological paradox that had prevented ChallengesЯus's first iteration to grow, and it remained an obstacle to scaling our current B2B operation. Of course, these were all thoughts and activities I threw myself into to keep up my façade of stoicism.

It was easier said than done to go full-throttle all the time, and it was often when we'd been in overdrive at ChallengesЯus for some time and I finally got an opportunity to breathe and reflect that I realized not much of real substance was coming back from all the sweat we were pouring in. In those moments, I had no choice but to question whether everything was unraveling. In those moments, I could become philosophical—spiritual, even.

Maybe we wanted success too badly. Maybe it didn't matter how hard we worked; maybe whatever was going to happen was predestined no matter what we did. Maybe the universe was trying to tell us something. Maybe the universe did not want to be dented, after all—at least not by us, or not the way we were doing it.

It was how I felt when the executives who had already collaborated with ChallengesЯus started disappearing, one at a time. Some executives were writing to say they were discontinuing our services, some were not responding to emails, others were leaving their companies under mysterious circumstances and falling off the map. One by one, our network of support was toppling like so many dominoes.

What was happening? Were they being promoted for delivering outstanding metrics to their superiors? I wished that were the case, but I knew it wasn't. I had learned from Magnus, the master statistician, not to extrapolate based on too few data points. One day though, Magnus brought the pattern up himself.

"This is way too many coincidences," he said. "The evidence is becoming way too compelling to be dismissed." We called the phenomenon The ChallengesЯus Malediction. We didn't need hard proof because we could read between the lines: our champions were being fired for choosing to work with us, and it didn't require a rocket scientist to understand why it was happening.

Anyone who has ever tried to sell something out of the mainstream to any major company, regardless of industry, knows how hard it can be. It certainly was in our case. The bigger and more successful a company was, the more it had to lose by trying something different. As a result, people in particularly large organizations tended to play it safe.[14] A manager's job is to manage resources in the most predictable and profitable manner possible. It is not to walk barefoot in the corridors wearing a black turtleneck sweater like Steve Jobs while telling people to "think differently."

Senior managers are often ultra-conformists. When confronted with the choice of buying Bain or Cole+Moore's consulting services, which are both dependent on manpower, or ChallengeЯus's unorthodox and unmanned change management gamified application, the average baby boomer decision-maker would typically opt for the former.[15]

As we saw it, our buyers had organizational headaches which, presumably, the IBMs of the world didn't have the creative vision to fix. Because those buyers wanted the best for their organizations, and they were working for tangible results rather than promotions, they chose us. Unfortunately, we theorized, it had backfired on them—big time.

What I found so ironic was that when employees did what they were actually *supposed* to do according to the official narrative of the company employing them (like finding solutions or being innovative and proactive), they ended up in trouble for refusing to play within the organizational bounds.

We had a few good laughs as we envisioned new alternative taglines for our company: "At challengesЯus.com, we help you execute on your strategy—but we don't just stop there. We also get you executed."

Still, we were learning a lot from our unexpected findings. One thing was that a great product and a great strategy were not nearly enough as a small company. Being successful as a new player also required someone on the other side to help, someone willing to stick their neck out. As strange as it sounded even to us, sometimes you needed cowboys in suits on the buyer's side, too.[16]

13. ALL HELL BREAKS LOOSE

After training for seven short weeks, the day of the marathon arrived. After speaking to Magnus, he had told me that a good marathon time was below four hours, which I took note of. The run began at 5 am from Kaapsehoop at an elevation of approximately 1400 meters.[1] The terrain was mountainous and wild, and at that time of the day, it was rather chilly.

We ran through the dawn, and the heat began to ratchet up as we made our way down the mountains and into the valley below. As the sun came up the rest of the way, the temperature must've been close to 40◦C.[2] It got hotter and hotter, and people around me started collapsing on the sides of the road, unable to continue. The pain and heat I was experiencing were oppressive, and the entire scene was surreal. All I could focus on was the finish line, and as my mind swam, it was hard not to draw parallels to the entrepreneurial journey. Many people wouldn't make it all the way to the end, but no matter what distractions came across my path, all that mattered was the finish line. When I finally passed the finish line, the tracker on my shoe registered my time. I looked up to see 3:55 in neon numbering on the scoreboard—not bad for a first-timer.

* * *

I had been trying to get in touch with Dennis, ever since he had left the healthcare company we were working with, but I had not been very successful. After months of unanswered emails and messages, we finally got back in touch. He was about to move from his native England to Belgium, where he had found himself a new HR role at a large chemicals company.

Through our time working together, we had connected on more of a personal level rather than a professional one. I was still very interested in validating our malediction hypothesis, and since we had a relationship of sorts, I wanted to bluntly ask him whether ChallengesЯus had been both the solution to his corporate problems and the source of his personal ones.[3] Even so, I thought better of it because I had a better idea.

On a phone conversation one day, I asked Dennis another question.

"Would you be interested in joining our board? We've had a fantastic working relationship, and we could use your strengths on our team," I said. "As you know, the work we're doing is cutting edge." With Dennis on our side, Magnus and I reasoned, we would not only stand to professionalize our board, but we would also have a better chance at accelerating our market penetration. Initially, Dennis didn't seem too interested—but he also wasn't a hard *no.* I sensed that with a little convincing, I might be able to talk him into it.

To begin the courting process, I stayed in touch with Dennis and we sent emails back and forth over the course of several months. Eventually, we decided to write a piece together about organizational change, which we got published in a major online human resources magazine. Our article was about gamifying work processes as a way of courting millennials. A few professionals in the HR industry were starting to talk about it, but very few were doing anything about it. ChallengesЯus was something of a pioneer in that area, and our article—although not a direct promotion for our company—showcased some examples of how a typical manager could think about engaging people from the bottom up by making work seem like play.

After developing our conversation with enough back and forth, the call I was waiting for finally came.

"I was thinking about this board position you talked to me about," Dennis said. "If that's still on the table, I would love to visit you and see where the magic happens, so to speak." *Wait until he sees what that really looks like,* I thought to myself.

"That's an excellent idea, Dennis!" I said. "We would

love to have you as our guest." After our conversation, I scheduled one of my now-regular business trips from Mozambique to Stockholm to meet Dennis. As I looked out of the only window of our small Kista office, a +44 number flashed on the phone display. It was Dennis. As I looked down, I saw him standing in front of the main door, waiting for me to let him in. I could not believe he had followed through—part of me assumed he would simply cancel at the last minute.

As I walked him through our office and introduced him to Magnus, Napster and Frida, I could sense he was not particularly comfortable. *Was it because he had gotten stuck in midday traffic on his way from the airport?* I wondered. *Perhaps it was the weather?* Though it was not unusual to lose 15C overnight during a Scandinavian summer, the swings that summer in 2016 were particularly severe.

We all sat down in the shared office's only large conference room (which, fortunately, was available). I struggled briefly to connect my Mac to the projector, but since Magnus and Dennis were chatting away, nobody really noticed. After making Dennis a cup of tea, we were ready to begin.

"I thought we could have you lead the conversation, Dennis," I said, handing him his tea in a ChallengesЯus mug. "I'm sure you have plenty of questions, so let's get cracking!"

To my surprise, the first thing Dennis wanted was a

short background recap about challengesЯus.com and a pitch as to why we saw him as an asset to our company. He listened courteously as Magnus and I went through it all, and every now and then, he would nod with approval. Often, though, he looked uncomfortable.

It was odd to have him across the table from our team in person on our home turf. I also realized it was the first time I had seen him in the flesh. Up until then, I had only come across his picture on LinkedIn and later on Skype. He was much shorter and thinner than I had imagined. It was strange to think that it had been about six years since we first got in touch.[4]

Dennis was a tough man with plenty of experience in senior positions, which was even clearer in person. I remembered how speechless I had been when he gave us the green light to use digital challenges to help engage his entire business unit. At the time, I didn't reveal that he had been our first client. Still, now I wanted to—his faith in us had let us come so far. It felt like overdue thanks, so I decided to remain quiet. Finally, we reached the end of our introduction.

"You've achieved all that with just the four of you?" Dennis asked.

"Yes," I replied. Dennis's bony lower jaw literally dropped, his eyes popping out of his head. From his reaction, I guessed that he would've preferred to hear or see something else—perhaps that we were a team of 50 computer geeks with oversized hoodies and Beats by Dre headsets. Maybe he wanted to hear that we had outsourced all the coding to some remote country full of underpaid developers. There was none of that at ChallengesЯus. What he was witnessing was the real deal, the command center where all the decisions were made. It was also from here that we had pushed a button and made more than 3,000 of his employees around the world do things they had never done before. To calm his rushing heart, I knew I had to come up with something that would appeal to his analytical brain or risk having him drop dead.

"It's marvelous Dennis, is it not?" I said. "Think how efficient we are to be able to do all this with such a lean team. And imagine how much better we could still be with your guidance as a board member." The elephant in the room had been addressed, and we quickly refocused on his possible involvement with us for the remaining time we had left together. After talking it through at length, it was time for us to part ways.

* * *

Seven—that was the number of near-death experiences we were supposed to expect before making it. At least, that's what Peo had told us early on, after one of our first real frights. As Peo explained then, he had gone through many such near-death experiences himself and knew many entrepreneurs who had as well. For our part, we had met plenty of people over the years who had given us similar accounts of struggle, pain and suffering before eventually finding a crack to squeeze through. Aram Sargsyan was one such example.

It had taken Aram more than 10 years to really get started with his IT security firm. At some point, maybe in year five or six, he had to let his staff go—all of them, 35 people. He kept only two associates and went back to square one—and after that, one of his two partners quit. Aram had to do consulting gigs for several years just to keep the lights on and figure out his way forward. Eventually he did, and ultimately, his company worked so well that it was bought by Cisco.

The same was true for Richard Branson. Branson's first business was *Student*, a youth-culture magazine run by and for students, which launched in 1966. *Student* was run on a shoestring budget and didn't last, but the experience taught Branson lessons that later let him give birth to Virgin. Virgin was not just a straight line of successes, either—many of its ventures failed over the years, including Virgin America, Virgin Megastore and Virgin Cola. It was a trajectory that could apply to any entrepreneur.

ChallengesЯus had already had a few close calls, but nothing compared to when we technically went bankrupt.

We had been burning our cash at a steady pace for a while and had even gone on the offensive with recruitments and marketing. We were taking advice that Peo had given us from early in the Sting process, when we were warned that it was possible to have too much money as a startup.

"If you have more than seven or eight months' cash flow sitting in your account, you are clearly not being aggressive enough," he had said. "You are a startup for God's sake, not the Federal Reserve."

Since we had gained confidence, we were now constantly operating with about four months of cash reserve —it felt unnecessarily stressful, but we were rebels. The reason we had gotten in such a cash-strapped situation was because we had needed more team members and now had six people on our payroll.

During recruitment, we had no idea about the kind of profiles we needed to look for. All we knew was that, as far as sales were concerned, we needed bad-ass closers because we had a model to prove. Although our investors had been quite relaxed for a while, we knew the time would soon come when they'd be breathing down our necks and urging us to make more of their money.

To increase the probability of providing Jonas and his friends a return on their investment, we had a two-pronged strategy: direct and indirect sales channels. As far as indirect sales were concerned, we had enough partnerships in place to draw some conclusions about what was working and what was not.

Out of around 10 partners we had enrolled, trained and certified, very few had delivered anything at all. A few had gotten us tiny pilots with some local companies, but most had not delivered anything of substance. The most promising pilot project we had at that stage was being run by a communications agency in the UK, a local partner led by another strong lady named Jo.

It was a one-year project for a small unit of one of the top five automobile brands in the world, and some 450 employees were taking part. Once again, we had had to burn the midnight oil for weeks in a row to customize our platform to meet the brand's stringent requirements; we had not had much choice. It was one of another in a long line of tough decisions: do something outside what we were trying to do, learn and get some cash in the process, or do nothing and get nothing.

This UK deal was not mind-blowing, but it felt good to be back in positive territory. What was exciting was that the client would most likely add many more users to the platform if they were happy with our services. From our perspective, that was worth a shot.

As far as direct sales were concerned, we had initially looked for senior salespeople with management consulting backgrounds and a genuine interest in tech. Truthfully, we wanted them to be like us: experienced and flexible, but above all, hungry and original. We struggled most with their appetite for risk.

There were inescapable risks in a startup—plenty of them. Our business could fold tomorrow or the day after. Being constantly shadowed by the fragility of life was not for everyone, and so we needed people who enjoyed living on the edge. All the team members also had to fit well with one another, and it wasn't something that could be faked or developed over time. No, in a tiny team, the symbiosis had to be instant and resolute.

To compensate for personal risk, we were offering company stock. We argued that the stock offered candidates the possibility of recovering short-term salary losses if we were successful in the long term. While it was relatively easy to convince inexperienced dreamers with such a proposition, it was virtually impossible to attract the type of sales-grinders we were after. It was how we realized that we had to lower our standards, and eventually we settled for greener versions of our dream candidates; even in recruitment, we were running tests.

Our latest was with Stefan, a 35-year-old man who was highly educated and motivated to join "the other side" after having spent years purchasing human resources solutions. There was also Sofia, a 24-year-old tech-passionate young woman with some frankly irrelevant consumer-product sales experience.

While Stefan was well-intentioned, it was clear that he was incapable of picking up the phone and leading prospects to the deal stage. He was way too soft and did not click very well with Frida. In a nutshell, he was far from a slam dunk.

As for Sofia, her lack of past exposure to the corporate world of tough decisions and constant cancellations was not making her days at ChallengesЯus a walk in the park. Even so, she tried very hard. Magnus and I liked her "can-do" attitude, and although she was not really cut out for her position, we kept her on for a bit longer than we should have. *Who knows*, we kept thinking, *maybe she could surprise us?*

Looking at our accounts receivable, our financial position looked under control since we also had money from our American manufacturing friends coming in. One day, however, Magnus called me 15 minutes after we had split up after lunch.

"Mayday, mayday," he said.

"Copy that, buddy," I replied. "What's up?"

"Secret Squirrel, we have a JDAM.[5] What is your 20?[6] Request for all personnel to return to base, over." Magnus was a big fan of the military. While I had done everything I could to avoid national service (which was compulsory in France back then), Magnus had volunteered for the toughest regiment one could possibly join in Sweden: the elite mountain infantry. As a result, he had taught me a few words used for radio transmission.

I cannot say I was passionate about it, but every now and then, we would "roger" this or that, just for the fun of it. As a result, I did not pay much attention to what he had said that day. In fact, by the time I reached the office, I had totally forgotten our cryptic exchange. As it turned out, it was a true "mayday" situation: we had now reached the edge of our financial cliff. Without obtaining the cash owed to us by our American customer immediately, we would be in deep trouble in mere weeks.

The money was already two months late, and yet,

despite frequent email reminders and calls, there was no sign that it was coming soon. Every time I managed to talk to someone, I got a different reason for the delayed payment: it was a bank holiday, it was the payment system, it had already been triggered and so on. At first, it all sounded very reasonable—but eventually, it was quite apparent that something was going on.

We wondered why such a serious company would do such a thing with its suppliers, but it eventually became obvious why. If a company approached its cash conversion creatively, it could be very good for its own financial position. Basically, one could ensure short collection times for receivables and long ones for payables. In the case of our client, there was not much we could do to remedy the situation or speed up payment. They had all the power, and we had no legal department to go to. It was a David and Goliath confrontation.

We realized we were just 15 days from technical bankruptcy. So, we put our last bullet into the barrel: the threat of a 10 percent weekly penalty based on the total amount invoiced.[7] Pointing to it and demonstrating what it would mean if we were to invoice for what would soon be three months of deferrals was enough for Goliath to reconsider playing hardball with us. We received our dues a week later—just in time to pay salaries.

We were coming to the painful realization that we were too "project-driven," as evidenced by our work with Dennis and all our other clients. After a company used us for one particular task or initiative, none of them could conceive of running "challenges" for other future projects they had—no matter how much we tried to convince them.

Regardless of how well we scored and irrespective of how pleased they were with our work, they wanted something different every time—something fresh to excite the troops. Meanwhile, smaller accounts were simply not profitable. We were dumping them one after another to focus our time and energy on where the money was.

From the outside looking in, we were doing reasonably well. We were attracting rather large customers and getting media coverage, we had secured investments, we had been finalists in a pan-Nordic pitching competition and so on. Even if we had some internal strategic problems, one thing we had mastered was the ability to position ourselves—to be seen where it mattered. It was why Bill Hardeman, the investor from the UK who sat on the Sting Day panel months earlier, had asked to meet with us.

We were excited for a chance to get another capital injection to keep things going. After a good hour of exchange, Bill spoke.

"I like you guys very much," he said. "I think you are great, I really do." We were not spring chickens, so we knew there was a "but" coming. "But I don't think you are scalable enough."

Had I heard those words years back, I would've thought, *yeah, yeah, we'll show you.* But we had already fallen flat on our faces multiple times. We had been shoveling filth for years, looking for an optimal product-market fit. We had tweaked our model and our approach so many times and had just narrowly avoided bankruptcy. It was frustrating, but by that point, I was so numb that Bill's words flew above me like the French Acrobatic Patrol above the Avenue des Champs-Élysées on the 14th of July.[8]

"Don't get me wrong," Bill said. "What you have achieved is, by any standard, remarkable. I think you have the potential to become a €100 million company, but it will take time. A lot of time. It will also be very labor-intensive, and your margins will probably be very small. That's not what we are after. We scout for unicorns, those few companies we can turn into billion-dollar valuations in just a few years." With that, we parted ways.

In the days that followed our meeting, a consolation prize was that one of Bill's portfolio companies had been making headlines—not because they IPO'd or were acquired but because they had gone belly-up. It made a big splash. In fact, it was one of the most humongous startup flops in recent history, connected to the latest trend of cryptocurrency.[9] Tens of millions of euros had gone up in smoke, and hundreds of employees had found themselves on the pavement. It was just another big VC bet turned sour, but it didn't matter; Bill still had many more candidates with the potential to offset his latest miss.

It was tempting at any stage to believe that more cash was a panacea, but we knew deep down that wasn't true.

Ideally, it was only when you had more demand than you could serve that an external capital injection was required, not before, so we still had some time left to wait.

* * *

Six months had passed since Dennis had visited us in Kista, and we still had not heard back about him joining our board. Claire, Chléa, Melvin and I had been in Mozambique the entire time, and I had work piled up to my ears. He must've been busy, we figured since we certainly were on our end.

Eventually, I got an email:

After thinking this opportunity through back and forth, I've concluded that I can't foresee myself adding much value to your business. I regret bringing you this news but wish you all the best for your future success. Many thanks for your consideration,

Dennis

He did not go into detail, and I did not feel like asking for more explanation. *Maybe he was right after all*, I figured. *Maybe there isn't much he could help us with*. In any case, we had way more pressing issues to deal with.

Magnus and I were in our fifth year of business, and we had not eased up our tempo one bit. If anything, we had never worked harder. On one level, touching so many employees' lives and changing some businesses for the better was rewarding. In about five years, we had likely influenced tens of thousands of people's lives. Similarly, by validating our assumptions and falsifying other theories, we were moving forward and learning things. The truth, though, was that we seldom stopped to look at things in that light. Even though a lot was happening, it felt that we were sinking in quicksand. We continued to press forward, looking for more success, but it was getting harder and harder to see any light at the end of the tunnel ahead.

Mo Gawdat[10] once said that happiness was equal to, or greater than, the difference between the events of one's life and one's expectations of how life should be. If he was right, then despite our successes, neither Magnus nor I were particularly happy. It wasn't because we had shareholders to satisfy; it was because we had higher expectations for ourselves.

Although it was now clear that I would not make the Forbes 40 under 40 list (as I had just celebrated my 40th birthday), we were still hoping to achieve Magnus's goal of running a company with hundreds of employees—but it was not looking good.

Either way, my lack of happiness didn't matter. All I cared about was that in the space of a year or so, we had gone from having reasonably good turnover as a result of our hard work to a sales pipeline that had been drying up—

a rather bleak forecast when it came to scaling our business.

We were trying everything we could to find the right model, but it seemed like the best option we had was to rely on the archaic phone-banging sales model. It was not what we had planned, but for lack of better alternatives, it was what we had been doing for the last year. And even that was not working.

Despite our many attempts post-Sofia (who we'd had to let go), none of the people we had tried to turn into sales machines were *Glengarry Glen Ross*–style sellers. But that was hardly a surprise. I knew only too well that our product was hard to sell. To be successful, one needed to have the right organization at the right time and be talking to the right person—but to get to talk to that person in the first place took determination, resilience and serious audacity. Undoubtedly, it was a skill set that took more than a few months' worth of experience.

The situation was bad—really bad. The only thing keeping our spirits high was Peo's advice ringing in our ears. At most, we had only had five near-death experiences as a company. Surely, we had a few more left to go.

14. A PLATFORM FOR GROWTH

Maputo, April 4, 2017

"How about 'forumidable?'" I offered with great enthusiasm.

"*Nej*, don't like," Magnus replied. "I have one: findr.com?"

"Let me check," I replied. After looking at godaddy.-com, I announced the news. "It's available!" As I leaned back in my chair, watching lightning bolts illuminate the dark African skies outside, I realized how late it was.

"Man, I'm dead," I said. "It's already 2 am here, and I have an 8 am tomorrow. I don't think we're going to nail this tonight, so let's sleep on it and see what happens."

"Sure, man," Magnus said. "Sleep tight and don't let the African froggies bite." After he hung up, I took off my headset and closed my laptop.

It had been about 15 days since Magnus and I started

our all-night brainstorming sessions; it reminded me of old times at Cole+Moore, when we first started spending so much time together, before we embarked on this insane journey.

We were looking for a new name that would better reflect our change in direction. Again.

It was our second major change in our short life as a company—and depending on how you looked at it, we were running our third company already. First, we had gone from a consumer app to a hybrid change management tool. Though we had earned ourselves more big clients that way, we couldn't find a way to get out of the middle of our processes. We also couldn't convert enough big customers into repeat clients and couldn't convince outside investors (or ourselves) that what we were doing was scalable. As a result, we had pivoted again,[1] and we were slowly but surely morphing into an internal communications platform.

Throughout the years, I had constantly examined all facets of our company, trying to propel us forward faster. It was neither about process improvement nor cutting excess fat; it was about extrapolation and seeing patterns where patterns were hard to see. It was about having a creative mindset, and it was where Magnus and I could get into heated arguments.

"This isn't happening fast enough," I would say. "Something must not be right. We have to be missing at least one key piece of this puzzle, somehow."

"Okay, but why jump off a building when it's clearly

not burning and there isn't even a hint of smoke?" Magnus would reply. More and more, we were looking at opposite sides of the same coin. Magnus was more focused on real observations while I wanted to go beyond what was right in front of us. At the end of these semi-monthly arguments, we always agreed to disagree. We could both be right and wrong at the same time, and ultimately we were dependent on one another to make things work.

While we never stopped believing that we were building the right tool to help organizations improve their own change initiatives, we also knew we had tried all we could to convince the market without the necessary success. Consequently, we had progressively warmed to the idea that given the way we were going, we might not make it.

On one level, it seemed that we were constantly changing our minds because we didn't know what we were doing. To some degree, that was undeniable; we had no certainties about anything. But who did? Was there even such a thing as being certain in the startup world?

It reminded me of a response former US Secretary of Defense Donald Rumsfeld gave in response to questions by the Department of Defense in February of 2002:[2]

> As we know, there are known knowns; there are things we know we know. We also know there are known unknowns; that is to say, we know there are some things we do not know. But there are also

unknown unknowns—the ones we don't know we don't know.

We had hunches and ideas aplenty, and we had tactical plans to try them out, but no one could know whether our vision would align with reality.

Sometimes I wondered if Magnus and I were coming up with half-baked solutions by meeting halfway more and more often. More frequently, I believed that we were doing what we had to do and going where the market was taking us. We were free-floating, in a way. Even if that sounded a bit New Agey—like something a hobbyist Alpaca herder rather than a team of funded software engineers and management consultants would say—it was true.

Adaptability was one of (if not *the*) primary "must-have" qualities for any startup. We were neither the strongest nor the most intelligent team in the world, but we were clearly demonstrating our ability to adapt to the environment and survive as a result. That adaptability was what led us to our third pivot, the process of which had begun about a year earlier.

It started when our UK partner had gotten a Request for Proposal (RFP) from an automobile company we had done a small project for. The aim was to develop an internal communication platform, and naturally, there were many strict requirements.

The platform would be the place where all employees of the UK company, some 4,500 of them, would go to find everything they needed. It was to offer internal HQ information[3] as well as groups where people could jump in and out on an as-needed basis. What they were asking for was a Facebook for corporations. They wanted it for an initial period of two years, and unsurprisingly, they wanted it fast.

It was miles away from what we were doing at ChallengesЯus in terms of business, but there was one thing we had going for us the other RFP recipients did not have: we had already done a pilot with the company. We had, therefore, already proved our worth and the stability of our platform, and we had also managed to convince at least some of the higher-ups in the organization that challenges were a tremendous way to trigger engagement in their teams. We had also proved that it could be done at scale. It was precisely why we knew we had to participate in the RFP.

At the beginning of the application process, our Oh-Crap-o-Meter had reached new levels. We only had three months' worth of cash left in the bank at best, and there was not much else to do. We had to lean into the discomfort and hope for the pendulum to swing our way. We were essentially asking for a miracle.

After we had sent in our application, we thought our answer to the RFP was decent. We had addressed all the company's needs and had created mock-ups that gave them an exact idea of how their ideal internal platform might look. Our pricing might have been a bit steep, but we had done that on purpose. We wanted to place ourselves on the high end of the spectrum.

It was a ballsy move, for sure, but we knew a company like theirs would never do the heavy lifting required to make that kind of product fly internally. They had neither the knowledge nor the ambition—and probably lacked the requisite human resources, too. Their job was to flood the market with cars, not to fiddle around with software.

What we had suggested went way beyond mere software. We were offering a plug-and-play solution. We were proposing to take over the entire process so they would not have to bother about anything at all. They would just have to pay the bills, and when a problem came up, we would fix it—somehow. We had no idea how we would do that in practice, but we had improvised before, so we were confident.[4] We also had the comms agency, who was our partner, on our side. They would deliver the actual service layer, and we would take care of the software part.

What we were essentially saying was that the client would not be able to do this project on their own and that even if they could, they shouldn't because they had far more mission-critical things to take care of. It was a bold statement, but it was true. It was our point of view, and we would stand behind it.

As I returned from my run that day, I had a few missed Skype calls from Jo, the communication boutique's managing director. I had left home unusually late that day because it had been pouring all morning. When it finally stopped, I took my chance. To get out of my head, I went for a long run.

With my marathon successfully behind me, I was now running for pleasure. It was hard to explain, but after all that training, running had become a part of my routine. I now considered myself a runner, and though I was not reading the running magazines you might find in the supermarket next to *Wired*, I was serious about it nonetheless.

I kept track of all my runs and of my times.[5] I carefully managed my recovery, and I ate properly (at least, better than I had before). At times, I even surprised myself by getting in passionate conversations with strangers about pace, cadence and past achievements. Though I was not following as crazy a regimen as some of my running partners (who completed several marathons a year and topped them off with an Ironman), I could feel myself getting addicted. Running, and the pain that came with it, was a drug. The more pain I endured, the more my body and mind demanded. It soon became a ritual reminding me that my pain threshold was much higher than I tended to think it was—and I was determined to figure out exactly how much I could take.

Because she had not been able to reach me, Jo had dropped me a line to tell me we were now one of only two companies left in the application process and that I needed to come down to the UK to pitch to the board with her. Technically, I would have to come "up," but Jo had no idea about the transcontinental commutes I had been taking for over two years. The leadership team wanted to see us the following day, but it would've been impossible for me to switch hemispheres that fast. After looking at flights, the best I could do was 48 hours.

* * *

When the meeting day came, I was impressed by the line-up of the executive leadership team. There was the group general manager, the global marketing director, the general manager for each brand (three in total), the head of sales for the group, the head of quality, the chief financial officer, the head of procurement, the head of legal and, of course, the human resources director. Meanwhile, I was exhausted. After my transcontinental flight, I could barely keep my eyes open, but I had to pretend I had only traveled a couple of hours.

After Jo had effectively set the scene, it was my turn to stand in the middle of that room and go through the slideshow I had prepared for the occasion. I had worked on my part of the presentation for two days. This time, I could no longer be a lawless cowboy. The stakes were too high. We had reached the end of the line. We either had to secure this deal and go for another round of crazy adventures or die in a ditch in just a few months. Our company was fully in the red, and it was all on me to get us back in the black.

After explaining our vision for their solution in my own words, I went on to answer specific questions they had. We were offering them an internal communications solution, but my plan was to talk more about business. I felt at ease talking about business because it was the world I came from. It was also what ChallengesЯus's strength had been all along.[6]

I knew, or at least imagined, that our last competitor in the process was likely a seasoned comms solution provider. As such, they would most likely approach the whole thing from a communications perspective, arguing for engagement metrics like number of logins, returning visits, page views and all that soft stuff. Of course those metrics were of interest, and we had tools to measure them, but I had to take their focus off those metrics. Our final competitor surely had an advantage in that area, so I was determined to push the conversation toward a narrative we could control.

I had looked up everyone in the room online before the meeting. All of them had spent their entire careers in the automobile industry. Some had been "car-hopping" their entire life, although most had been with the same employer for 20 or even 30 years.

They were not creatives coming from Disney or tech-inclined folks who worked for Apple or Tesla. All of them, besides the head of HR, were guys in their mid-50s who wore dark gray suits. They were an old-school conservative bunch in a conservative industry who were selling cars, not Tamagotchis. What they all had in common was they had all made their way up within the industry itself. They knew nothing else.

My hope was that feel-good vanity metrics like comments on articles and number of likes would not matter as much to them. To get them to invest in a tool like ours, I would have to demonstrate clear value for their money, down to the cent.

As I talked about gamification and how we had designed the platform to motivate genuine employee engagement—one of the key things they were after, at least as per the brief—the group general manager literally snarled at me.

"I don't believe in your gamification or whatever the hell you're talking about," he said, his arms crossed over his inflated chest to demonstrate his authority. He was clearly the alpha male in the room; not surprisingly, he was sitting dead in the middle of a U-shaped arrangement they had made with the chairs and desks. With his words still ringing in the air, the others suited men turned toward him and nodded in almost perfect unity. It felt like I was watching a synchronized swimming gala but without the music or the eye-pleasing athletes. Finally, all the men rotated their heads 90 degrees back toward me at once. Twenty balls of clear gel were staring back at me and expecting, I guessed, a well-articulated counter to their boss's verbal thrust.

I had few options: either I let myself be run through once and thereafter risked having to systematically defend each and every one of my points, or I went on the offensive, stood my ground and refused to be ostracized. In doing so, I could win the respect of the group and, perhaps, even the alpha.

"I think I understand where you're coming from," I said with determination. "May I suggest we go back one step?"

He looked at me with an unmistakable air of superiority, not even feeling the need to answer my question. The rest of the panel watched intently. They were ready for the confrontation.

"If I may, sir, I think we believe in the same thing. We just seem to differ on the semantics," I began. "Let me put it this way: can we agree that we have to have some way of making sure that people, in this case your employees, do what you and all the managers down the org chart want them to do?"

He waggled in agreement, much like a silverback gorilla would. He was not yet with me, but he had at least uncrossed his arms and now had his hands joined in front of him, elbows on the table. He was moving his two index fingers from the top of his nose down to the tip of it, repeating that same movement, over and over. He was ready to dissect every single word I said.

"To enable that, we have designed features within our platform—call them levers or mechanisms, whichever you prefer—that will motivate, reward and secure those exact behaviors we, or shall I say you, are looking for. These can be anything you want for your employees—for example, do more of this or do less of that. You name it."

"Are you talking about change management?" he said.

"Yes! Yes, that's precisely what I'm talking about: the incentivization of sought-after behaviors."

"Now, that's interesting. Tell me more!" He said to his colleagues' surprise and my complete astonishment.

"Well, to be very concrete, we have tested many things over the years. Today, we know *precisely* how to attract people to the platform—regularly, not just once—and to take them where we—or, again, you—want them to go. We know how to trigger interest. We know how to create a sense of urgency. We know how to sustain enthusiasm, and so forth. And we know it works since we are doing it with some of the world's most celebrated organizations."

"I see," he said, now completely absorbed. I needed something powerful and preferably daring to send my presentation home. Sensing that he had fully bought in on my change management proposition and knowing that I had to differentiate our offer from that of our competitors, I went bananas.

"You see, at the end of the day, 'change' is what we do. It's what we've done for more than 10 years," I said. "If I'm perfectly honest with you, I personally could not care less about 'likes' and all this yuppie nonsense. They are mere gimmicks we leverage to align with our audience. That's what younger folks want these days. Trust me, with *that* audience, you can't talk about stages of behavioral change. You can't use Lewin's model[7] of 'unfreeze, change, refreeze.' Instead, it needs to feel easy. It needs to feel like a game, with virtual rewards, points, medals and all that shiny stuff: gamification or behavioral change. It's all the same stuff, really; it's just different packaging."

"You know what, I'm starting to like this 'gamification' idea of yours. Thank you for the clarification." As soon as he said it, I knew I had him. After that, Jo and I left the meeting and made our way to the reception area to leave our nametags.

"You were quite gutsy in there," she said in proper British English, with a little irritation in her voice. "Now, getting the deal is going to boil down to whether or not 'the Man' swallowed your rather mouthy sermon in there."

"He did," I replied curtly as we stepped onto the elevator.

* * *

Waiting for the firm's decision was simply excruciating. Despite the self-assurance I had exhibited in London when Jo and I parted, I was now in agony.

I had been home for over a week and had still had not heard a thing—not even a "thank you for coming." We were one decision away from the end of our little adventure or whatever came next. We had no idea which way it would go, what their decision criteria were or how we might have scored on them. All we could do was sit, wait and take comfort in the fact that we had given it our all—and not just in the pitch. We had worked so hard for so many years. It was now all up to the universe.

As much as I tried to remain positive,[8] I had a hard time repelling the dark ideas that kept creeping into my mind.

The fabled fat lady Magnus had referred to so many times over the years was clearing her throat and getting ready for her performance. We, too, were becoming a mere data point on the plot of failed startups, another casualty of creative destruction.

Adding to my stress were Caroline's phone calls demanding updates on how we were going to turn everything around and Klas's prodding about how I would turn his investment into profit. I felt sick. I could practically see our company logo hanging on Henrik's wall but as a cautionary tale. We would be one of the companies he forever mentioned to ambitious youngsters visiting him in his inhospitable basement for a condescending lecture about the pitfalls of entrepreneurship.

These bad trips luckily ceased the moment we got the email—the one we had all been waiting for. Though we couldn't believe it, we had somehow emerged victorious! I was so relieved that I was practically in tears at the news. It didn't matter that our newest massive win meant we were entering a new industry, one that was totally alien to us. It meant that it wasn't all for nothing. We were getting another shot.

Although we knew nothing about "internal communication," we now knew the drill. We would simply continue to do what we had been so good at doing all along: market the shit out of ourselves. We would blast everywhere that we were now the sole provider of internal communications services for one of the biggest automobile groups in the world. With the million Swedish kroner we would be invoicing per year[9] for two years (the length of our contract). It would give us some time to figure everything out. Magnus, the kids and I were on top of the world again—and so were Caroline, Jonas and Klas.

Everything was finally coming together. Through all our learning by doing, trying and failing, we had become what we were destined to be: an incredible internal communications tool! It had taken us more than five years, but we had figured it out.

Over the next few months, we did a lot of things both internally and externally. We zeroed in on our new name, and the necessary changes cascaded down throughout our operations. We were now going to be xtrnlDesk, which I liked. It sounded edgy, it felt techie and it seemed to convey the business we were in.

Unlike a typical comms service, which by design is a function of the organization, we were offering the same service as an outsider. We would not only provide the tech, something most companies were already outsourcing, we would also offer, through our partnerships with agencies like Jo's, an admin and curation layer on top of it.

It was a good deal for clients since it was much cheaper than having those capabilities in-house. The beauty of our promising model was under the hood, though. We were essentially the first startup ever to offer a mass-customizable platform like Facebook for corporate use. It felt fresh and innovative. Thanks to our modular tool, we could have customers decide which features they wanted and activate them with the flick of a switch. It was a way for the whole team to get out of the middle of every interaction. This was something we could scale.

We also went on to develop the first version of our app for Android and iPhone. The activity in the office was crazy, and there was so much enthusiasm. Meanwhile, I continued my transnational commutes with renewed vigor. Almost half the money I was making per month was going towards plane tickets, but the flights allowed me to enjoy the best of both worlds. Still, I never got used to the extremes.

Our customer acquisition with the new model was still sluggish despite the many things we were doing. Sure enough, one year after our big win, it was clear that we were still not capable of flying on our own. To have a fighting chance to break through, we needed more cash—a lot more.

15. THE PARTY IS OVER

The first thing we did when looking for new money was to figure out where it had the highest probability of being found—and that was not by knocking on Bill Hardeman's door. No, our opportunity was on the fourth floor of a building in central Stockholm. We were going back to ask Klas for additional funding.

This time, we came highly prepared and stepped into the meeting room with complete conviction: we would take over the world with our software, one communication platform at a time. After all the babble and hyperbole, Klas, Jonas, Jan and Per were ready to hear our projections.

We had spent weeks crystallizing our pitch, and in the end, we had come up with three different possibilities, all of which had merit and showed the impact we stood to make based on different investments. With very little money, we would be quite conservative in our approach; we would add a few people to the payroll, but overall, it would be slow and, frankly, quite boring. In our view, this strategy was a non-option.

At the opposite end of the scale—that is, with a massive investment—we would be very aggressive. While we would be able to achieve more that way, the distance between the most aggressive option and the one in the middle was not particularly cost-effective for them as investors. At least that's how we framed it.

In doing so, we judiciously positioned our middle option as the sweet spot: a moderate investment with good potential for growth. It was how people tended to make decisions when faced with similar choices. When going to a restaurant, people usually didn't go for the cheapest dish, nor the most expensive. Instead, they chose something that looked like it would be light on the tongue without being too heavy on the wallet (which restaurants knew as well, of course).

With our middle option, we would be able to pursue our tech development roadmap to acquire the expertise we lacked and allocate enough money to market ourselves.[1]

Juxtaposing our two investment meetings, the initial one where we first met Jonas, Klas and partners and this one, it was clear how far we had come as a startup team—we were now manipulating concepts like pre- and post-money valuations[2] with ease.

Klas and his colleagues were not stupid. They knew we would come back to our sugar daddies asking for more money. Interestingly enough, we didn't have to fight them to get what we came for. The truth was they were impressed with how far we had gone with so little money, and it was clear that we had been very careful.

A problem often reported by investors funding first-time entrepreneurs is that complacency tended to set in quickly. Eagerness and modesty at the outset rapidly faded, leaving a vacuum for hubris and sloppiness to creep in. All too often it leads to the inescapable demise of the team and project. Fortunately, we had demonstrated that we were the antithesis of that scenario.

Hearing them talk about investments, it felt like we had moved up to the top-right quadrant of the funding matrix.[3] I thought about this because it felt these guys were merely making gut decisions. They were real humans and even a bit cowboy-ish themselves, unlike so many investors we had met at private equity firms who only considered projects through the lens of an Excel spreadsheet.

As we quickly discovered, leaving the world of McKinseys, Bains and Cole+Moores to enter this new market was not going to make our lives easier. We were now swimming in a pond with some monstrous fish. We were battling the kings of internal communication: the IBMs, the Microsofts, the Jives, the Slacks and so many others.

Going head-to-head with these legacy systems was hard, but at least it seemed possible. They had plenty of money in the bank and a lot of resources, but internal communications was more a side gig for many of them rather than their core business.

Just as Tony Robbins and so many Shaolin monks before him suggested, energy flows where the attention goes. It also seemed that money flowed where attention went, so that would be our angle. We would go all in with our marketing money. Positioning ourselves against the big boys would, in some ways, be familiar. We had done it once already when we went head-to-head with the Big Four consulting companies. We had fresh ideas, the latest tech and we were nimble and creative. Where it took two nanoseconds for us to decide whether to do X or Y, it probably took months for our bigger competitors to make an equivalent decision. They had to fight internal bureaucracy, wrestle through red tape and tiptoe around egos and rigid bosses.

For many years, we had been playing with virality. In fact, it was probably the biggest question we'd been trying to solve internally since we started our company. It was a tricky topic, and I was obsessed with it. When we were still in our B2C[4] business, the main question was: "What would drive people to join, participate, comment and like?" We had naturally started on the tech side of things, that is with features, mechanisms, user experience and interface—the UX and UI as they are known in the trade. Though those are technical elements, there was still an undeniable human component to them. At the end of the day, they still started and ended with people.

Napster and I often clashed on that topic. Though we agreed on things that had to be built, he looked at them from an engineer's perspective while I came from an end-user one. For example, if you pressed a button or swiped left on your phone and got what you were supposed to get as a result, Napster would say, "You see, it works!" From my perspective, it might have worked, but the aesthetic was missing. It would probably have been a hailed output if it had come out of the Xerox Palo Alto Research Centre back in the mid-'70s, but not in present-day Kista.

Aesthetics, in my opinion, had become even more relevant now that we had shifted to internal comms. All our competitors had clunky systems with architectures built on old-school folders and on-premises software. The end-user experience was so dreadful—so bad, in fact, that clients were forced to spend millions every year training their employees to use the software. It was a juicy business for the providers but a terrible deal for customers.

Magnus and I wanted our tool to be as easy to use as any consumer-driven app and as beautiful to look at as the iPhone or the Nest thermostat. In fact, we were arguing that there should be no distinction in usability and user-friendliness between an app used as a private person and one used as an employee. Why should there be? I was doing the Steve Jobs thing, obsessing about the purity of the design. Other days, I was paraphrasing Einstein: "Let's not develop a simpler product, but let's make it as simple as possible." It became our internal development philosophy and our differentiation strategy: simple but not simplistic.

Over the years, our company had written countless articles on the topics of virality, of starting and sustaining an internal movement, of "crossing the chasm"[5] inside organizations, of finding true ambassadors and detractors and so much more. We were sharing all we did in real-time on social media to find and cultivate our own audience and build a community of aficionados.

We tried a lot of different things: short posts, long posts, super-smart posts, funny posts, posts with pictures, posts without pictures, posts with videos and every other sort of post. We undertook and published our own research, and eventually, we came to witness a similar pattern between "internal virality"—that is, how ideas flowed within closed networks and organizations—and "external virality" in open networks such as LinkedIn. We even developed our own mathematical equation, which we leveraged in sales conversations to demonstrate our deep understanding of the issue at hand. With this, we could almost predict the success of any internal campaign a client would have on our platform.

Still, despite burning large sums of money on promoting our own content on media platforms such as LinkedIn and Twitter, virality was biting us in the ass. Our return on investment on ads was far from awesome, and our content was not being spread all over the Internet. Very few people were cheering for us. Our conversion rates on ads were so minuscule that, at times, we started thinking we were the source of the problem. Our targeted buyers were not interested in what we had to say at all.

The truth, though, was that things were going just as they were supposed to. Typically, conversion rates tended to be bad for everybody. In a way, it was very much in line with anyone's experience on Facebook or LinkedIn; initially, you may notice the one ad in your feed, but eventually, when there are thousands and thousands of them, you no longer pay attention. You simply scroll by. The parallel with Bangkok's superhighways was suddenly so apparent—there were just too many businesses at any given time flashing billboards to the same group of potential buyers. And every day that went by, breaking through the clutter became trickier and trickier.

None of it, however, compared to when Facebook officially announced that after many successful pilot projects with their own employees and select clients, that they were going to release their own platform for internal communications.

"Holy shit!" I yelled when I first read about it. It felt like the sky had fallen. For almost a year, we had been telling prospective clients who had a hard time understanding what the heck a "social intranet" was that the easiest way to think about it was to think of "Facebook but for work." We had been gradually warming up plenty of big corporations to the idea, and then *boom*, there it was: the *real* Facebook at work was about to launch with Workplace from Facebook. At some point in our development, we had stopped trying to take down Facebook; now, Facebook was coming to take us down.

This whole situation felt outright unfair. Because their announcement had been so unexpected, business publications were writing countless articles about it, and they were being eaten up by people all around the world. It was not just the main editorials that were talking about Workplace. *Everybody* had an opinion. Some people hated the idea, but most people loved it—especially the people we were trying to court, namely, the heads of internal communication.

In all, xtrnlDesk had managed to enroll around 15 third-party partners who were all working hard to push our solution. Of course, Facebook had considered going through the same channels we had, meaning finding third-party partners who could help them "bring employees together." Overnight, hundreds of comms agencies, IT integrators and software development firms of all shapes and sizes had raised their hands to get a chance to work with Facebook. Even the kinds of huge businesses we had been chatting with—the Deloittes and the Publicises—had decided to go over to "the dark side." On our side, there was very little movement. It felt like I was chasing people down like a common street dog.

The one silver lining was that from now on, Facebook would do all the legwork for the whole industry, selling the idea to the corporate world that internal communication platforms of the future would be "social" and "in the cloud." That was certainly a good thing, as it would save me a lot of time trying to explain the rationale and advantages behind such an approach. Secondly, we knew that

Workplace, no matter how glorified its upcoming debut, would start with a heavy burden of its own.

Undoubtedly, there would be plenty of hesitation on the part of corporate buyers to deploy something that looked like a social network in their own organizations; for the most part, employers had been trying to ban the use of social networks at work. Most importantly, though, there was a general worry about how Facebook would use employee data. After all, they were not exactly in the public's good book when it came to privacy and business ethics. For those reasons, we concluded, not all companies would go with Facebook. Ultimately, there should still be enough for us to chew on after the Workplace tornado had passed.[6]

In retrospect, we had been synthesizing too much happiness. Just like any other human experience like anger, joy, hunger or sadness, happiness can essentially be fabricated. Research has shown that the experience of happiness is largely influenced by one's ability to turn seemingly unfavorable situations into positive ones. As Dan Gilbert, professor of psychology at Harvard University, puts it, "Natural happiness is what we get when we get what we wanted, and synthetic happiness is what we make when we don't get what we wanted." As it turned out, despite an overabundance of commonsense arguments against Facebook, many were still dazzled by the new shiny object that Workplace was.

I had always been told that in business, it was about the

product, not the strategy. It made sense to me. There were products I knew that had not achieved market dominance or even made the slightest impact on anyone's life, all despite their superior design or technical brilliance. Similarly, the market was flooded with poor gizmos and terrible companies that broke through and maintained their positions thanks to masterfully executed plans.

While this was true in every domain, I had particularly seen it in the change and internal comms platform business. When our company first ventured into the industry, Yammer was embedded into Microsoft's Office 365. While the product had been technologically superior before being acquired by Microsoft, by the time we were in competition in 2017, most players (including us, an insignificant tech company out of Stockholm) had caught up with the market leaders.

There was, for instance, very little innovation coming out of the Redmond, Washington–based Microsoft Corporation, and yet Yammer was on the lips of every communications director. We were showing them an interface that was a hundredfold more engaging (in my opinion, at least) and were demonstrating with real data the stickiness[7] of our tool compared to theirs. But despite all our data, nothing would change our interlocutors' minds.

The only reason for it was that Yammer had become a built-in extension of the Microsoft Suite, and as such, any corporation running Microsoft—85 percent of the world's organizations—had access to it for free as part of their license. It was a suboptimal product with a brilliant strategy. Buyers knew there were better alternatives on the market, but no one ever got fired for using a Microsoft product, especially when it was free. When Facebook joined the party, however, the novelty effect and the massive organization behind it made people wonder whether Workplace wasn't an internal comms savior. Of course, it was not; it was just another PYT.[8]

With limited gunpowder, we were shooting off the argument that companies and leaders should know better. They had tried Yammer. They did not need yet another "me too" product for an issue as complex as internal collaboration among employees. They needed a customized product with a service layer on top, just like the one we were successfully running for several companies out there (now including a top automobile brand). But it didn't matter what we tried to say. We simply could not put our money where our mouth was.

This time, our investors turned a deaf ear as we went back to them, begging for cash to compete. Again, we tried to convince them that "not throwing good money after bad" was a fallacy and a misconception. The fact was that we had worked our asses off for well over half a decade and had tried so many things. Surely, and sooner rather than later, we would nail it. The answer we received was, "Maybe, but not with our money."

It was expected. They were rational agents, and even if they liked us, they knew they should never put irrecoverable investments into an account following any "sunk" costs. Instead, they offered us a deal. They would put more cash in the system if we managed to find another investor who would match them. We were good at selling and had a lot to show for ourselves, but we also knew that no one in their right mind would go for that. Why would they? When your lead investor demands that someone else carry part of the cost, it doesn't send the best signal.

We tried for a while to look for dumb investors but couldn't find any. Eventually, we were spending more time looking for finances than looking for clients, and as we did that, our cash reserves were being depleted. It was how we concluded that we could not successfully raise a new round. At best, we would be able to do a "down round,"[9] which is never a good idea. Down rounds slam both valuations and spirits.

We had come full circle. In 2012, we had been determined to take down Facebook. We were looking for every possible opportunity to leverage the social network to position our anti-couch potato movement with challenges. Now, six years later, the hunter had become the prey. Thanks to Facebook, Workplace could literally influence billions of people by placing ads in everyone's feeds at a discount.[10]

By 2018, Workplace had made good progress, and at this time of writing, it is leading the enterprise social network market. We had fallen behind—way behind. Even so, we had managed to survive the five-year mark, which Anders had said only 20 percent of companies worldwide managed to do. All the same, we were now on our knees. The gentle melody in the background, the one I had heard on and off throughout the last year, was now loud and clear. It was the fat lady, and she was having her moment at last. The party was over. There was no longer any doubt about it.

We never learned whether Peo was right in his theory that seven was the magic number of near-death experiences one should survive to make it to startup nirvana. All we knew was that we had run out of steam during the last stretch. After more than six years of roller-coaster emotions, from anxiety to euphoria, bliss to desolation, we collectively chose to throw in the towel. Of every challenge we ever went through, it was the hardest thing we ever did as a startup.

16. THE LAST PIVOT

Stockholm, January 23, 2019

I am sitting in the waiting room of Honestlii, a Finnish startup that is stealing all the recent headlines, looking at a framed quote right across me. It reads, "This too shall pass —Abraham Lincoln."[1] The irony isn't lost on me. *I guess it will*, I think to myself. I'm nervously waiting for my interview with Chris, the chief revenue officer at Honestlii, trying to think through the questions I will soon face. My résumé of the past few years is a little unorthodox, after all.

After the whole Workplace by Facebook episode and our inability to raise further cash, we decided to begin dismantling our operation. Unfortunately, because we still have commitments to a handful of clients, we have since become a zombie company. We have stopped seeking new investments and pouring more money into developing our platform, but we still need to see our current contracts through until the end.

In the meantime, the kids have returned to RedOnion and now spend 100 percent of their time and energy building websites and mobile applications for small and medium-sized Swedish companies. They are also now married, which makes me happy.

Magnus is leaving the startup world to run some large IT projects for the Swedish police. I ask him about it sometimes, but he says that if he actually tells me all the details, he will "have to kill me." He and his family are now enjoying a peaceful life outside of Uppsala. I am the last person remaining, facing an uncertain future. Disillusionment aside, I must get past those feelings and start making money again. More importantly, I must get back into the swing of life, something meaningful. The best I have come up with so far is to interview at Honestlii. They run custom employee engagement surveys through an app and are trendy, which seems like a natural extension given my experience and skills. Now, I'm sitting in their front office.

A moment later, the door opens.

"Arnaud?" Chris asks. Chris is a tall, thin man sporting a bun haircut and a long yet perfectly trimmed beard—the trendy hipster look at its finest. He is walking toward me at a calm pace that feels almost exaggerated. He seems relaxed for someone who runs the corridors, but then again, of course he is; he doesn't have to impress me. My heart was racing before, but now it is in overdrive. I stand up from my chair with a big smile and shake Chris's hand as we make our way to the meeting room

It only takes us a moment to get there, but in that time, thousands of convoluted thoughts race through my mind. How much can I explain about ChallengesЯus? I hoped for it to be something that changed the world, but it ended in failure. As much as I am now trying to keep up a brave face and brush off the last six years, it says a certain something about me for having tried and failed, does it not? But to Chris, does it say that I am a brave entrepreneur destined to take risks, fight against all odds and innovate or that I am an unpredictable, naïve and erratic person who doesn't fit into corporate *or* entrepreneurial life? As I look at Chris's face in the moments before his first question, I try to glean as much information as I can (which is not much).

"So, Arnaud," Chris says. "Your résumé shows that you are a founder yourself! ChallengesЯus looks like an interesting venture, and from what I've seen online, it still seems to be operational. It begs the question: why are you interviewing to be a part of Honestlii?"

Here it is, the moment of truth. The good news is Jo, our UK partner, has already approached me about acquiring ChallengesЯus's platform and assets after the final client contracts expire. The first figures she threw out were devastating to my ego, but it will at least be a small consolation prize if the deal goes through. Her idea is for her and her team to use our platform as part of the newly created digital arm of her communications agency. Of all the possible fates

ChallengesЯus could have, it certainly isn't the worst one. It also means I have a graceful answer to Chris's question.

"Well," I begin, "The adventure we've been on with ChallengesЯus has been incredible, but I'm looking to make an exit. We are currently in the process of being acquired, which we are very excited about." At this, Chris's eyes brighten.

"Congratulations!" Chris says. "Please, tell me more—are you satisfied with the offer?" I consider keeping up the ruse that the company is doing exceedingly well and that I just want to take the money and continue with my life, but something about being in an interview setting like this again changes my mind. I notice a shelf on the wall behind Chris, and it takes me back to Henrik's basement. I am now picturing myself as one of Chris's trinkets: small, decorative, a footnote in the larger story of Honestlii's office. I realize I can't give the false speech of inspiring success that I planned to give to secure this position. I am too tired.

"Honestly, Chris,"[2] I say, "I have to come clean. I am not satisfied with the terms of the deal at all. I'm going to take it eventually, but it pains me to even think about selling ChallengesЯus. But I'm afraid I have no other option." At this, Chris frowns slightly and seems confused.

"What do you mean?" Chris asks. As soon as the words are in the air, the room already seems so much less heavy. The knot in my chest is loosening and I exhale. After that, I tell Chris everything—the complete, unvarnished truth.

I explain our early struggles to get into Sting, our

naivety about how easy it would be to run a tech company without a tech background, our struggles to raise funding, our multiple pivots, our modest successes, the guerrilla marketing strategies we poured our entire hearts into, our shared passions for changing the world and making it a better place—and how all of it had ended because we never found a way to package our product and our vision into a working business model.

Watching his face, it is clear this is not what Chris expected me to say in this interview.

"I'm surprised," Chris says. *This interview is over*, I think. *Honestlii wants to hire a winner, not a failure*. Still, after the pain of the last year, I'm too numb to care. There will be other interviews and other startups in my near future—there are always other interviews and other startups. I am about to get out of my chair to show myself out when Chris says: "It sounds like you have learned not just how to run one startup but perhaps how to run two or three. What kinds of skills have you taken away from all these experiences?" I freeze, hardly able to believe what I'm hearing.

"I have certainly learned a lot, though I was not entirely prepared to speak on all this today!" I say. "I know that if I had it to do over again, I would have done many things differently."

"Like what?" Chris asks. *Maybe this isn't over after all*, I think. Again, I tell Chris everything—what I wish I had known about MVPs, about traction vs. scalability and every other entrepreneurial principle in between. In return, and to my surprise, Chris starts telling me about the struggles Honestlii is having, and I find myself listening with intense interest. Our interview is supposed to last 30 minutes, but by the time we finish, we have talked for more than an hour. Finally, it's time to part ways.

"Let's do it like this," Chris says. "This was a fascinating interview, and as a next step, I'd love you to meet with my CEO. I'm sure we can make room for someone like you at Honestlii."

"Fantastic," I say as we shake hands.

"You know," Chris says, "ChallengesЯus may not have gone the way you wanted it to, but you just told me a hell of a story." I smile sheepishly.

"Well, thanks for saying so," I reply.

"I'm serious," Chris says insistently. "If that entire story were written down in a book, I'd read it." With that, Chris walks me back into the lobby where we set a follow-up appointment.

I leave the office in the same elevator I came up in, replaying the entire spiel I just gave from the beginning, trying to figure out what came over me to put all my cards on the table like I did. As I step out of the building and walk back out into the bright streets of central Stockholm, I feel a smile on my face. *This is just another in a long line of last-second pivots,* I think to myself. *I wonder where this one will lead.*

EPILOGUE

I can't lie—at first, the collapse of what I thought might become my life's work was not an awesome learning experience, and I did not absorb any lessons from it with grace and enthusiasm. I simply did not. It had to sink in for a while.

For over half a year, I told myself that I should not think about ChallengesЯus at all. I wanted to do something else, though I was not exactly sure what. I knew I had to adapt and overcome, just like the Navy SEALs often preached, but I didn't know where to put my energy. It was only after telling my entire story a few times that I had an idea (outside of finding a new short-term way to make income) of what I would do next.

I remember when I was about 14, a literature teacher told me that I was not a bad writer. I thought nothing of it. Over the years, like many of us, I came to write a few things. I wrote countless assignments at school as well as a master's thesis (which was not easy). At Cole+Moore, I wrote reports. At challengesЯus.com, I got a few articles published in magazines specializing in topics like change management, innovation and internal communications. I always enjoyed writing them, and it never felt like a chore. It made sense that at some point I would turn to writing.

At first I wrote for myself, as a sort of personal therapy. By the time I had finished compiling all my notes from this entrepreneurial adventure, though, it became evident that the questions Magnus and I had been asking ourselves all along were not unique. Millions of people were probably also asking themselves those very same questions. Next thing I knew, I was convinced that I could turn what I had learned into a book—not for myself but to serve others. To genuinely add value to others, however, I would have to open myself up—for real. I would have to retell how it truly unfolded and say what I truly believed. I would have to muzzle my ego (a hard task for entrepreneurs, as they tend to have rather grand views of themselves), but I knew it was what I wanted to do.

To my complete astonishment, the book that is now *How Hard Can It Be* came to me almost effortlessly and without writer's block. I just wrote. The process was a kind of personal development journey where I tried to understand what had happened to me, why we had failed despite seemingly having it all on paper (and to a certain extent, beyond the paper).

I initially thought it came to me so quickly because I

had lived the story. It was my story, after all, so it was easy to recount—but was that the real reason? As I looked more closely, I saw that the story was not my story at all—it belonged to Magnus, Frida, David, Jo, Peo, Henrik, Claire and everyone else who took a leap into the unknown. As I reflected on a truly incredible chapter of my life, I realized that I'd had the good fortune to work with some of the most hardworking, smart, resilient and passionate people I have ever met. I learned so much with them and from them, and the result was that whatever insecurities or shame I felt was replaced with gratitude (well, maybe not all of it, but most). After the gratitude came a new swell of confidence; I was getting my mojo back. And soon it was time for my next wild idea.

What if the seed of what we should do and be was always inside us from the beginning? What if it had always been there, obscured by programming from the external world? What if that was what the game of life was really about—to simply understand what that inner seed we had forgotten really was? Upon asking myself these questions, I thought of my life as a game of subtraction, a gradual peeling away of beliefs and constructs that no longer served me.

As all my ideas were making it into the book, I knew I had to share them—so guess what I did? I went to Jonas, of course! It was a bold move, even for me. At first, Jonas was skeptical (to put it generously) of a book idea that put the spotlight on failure with my startup as the subject. But I did not give up. A few months later, I had convinced my investment cowboys that we should get back on the horse—that he, Klas, Caroline and the others should help me finance the marketing of *How Hard Can It Be*. Just like that, I had moved from the depression end of the emotional cycle back to the optimism of a new beginning.

My confidence was not coming from the success of my startup but from its failure, or more accurately, from knowing deep inside that my failure was pushing me forward. In a sense, the journey ends now with you reading these words: although only time will tell what new beginnings this ending will lead to.

As I've learned, what we commonly call "failure" is only true failure if we don't grow from it, if we do not move forward as a result. Failure might just as well be called feedback—feedback from the environment. In a sense, it is the universe saying, "Nope, this way doesn't work. Try something else—if you dare."

For myself, and for anyone who can relate who reads this, it leads to another question: what's next? What happens when you set out to dent the universe and realize you've barely even bruised it despite giving it your all? The answer is as simple as it is potentially unsatisfying: you pick yourself up and you start over—just like you did on your way there. Just like you've always done. The outcome is never certain, but what is certain is that we all must fail our way to success. This book is just the latest iteration of that never-ending cycle.

My wife, Claire, is still saving the world (though for now, it is thankfully back in Stockholm). Due to COVID-19, she and I get to spend much more time together because we both work from home. Our family has grown, too, with our son Tim being the latest and final addition to our little team. Is there a greater plan at play? I'm not sure. Is this a Disney-style ending? I don't know. I only know that this time, it won't be about the trophy because there may never be a trophy. It will be about the journey: because no matter how hard that journey can be, all you can do is enjoy it. Whether you bottom out or fly, there is never a finish line.

If you enjoyed *How Hard Can It Be*, please leave your honest review on Amazon or drop some kind words on your preferred social media platform!

NOTES

1. Going Live

1. 'The invisible hand', introduced by Scottish philosopher and economist Adam Smith in his book The Theory of Moral Sentiments, in the 18th-century, is a metaphor that characterizes the mechanisms through which beneficial social and economic outcomes may arise from the accumulated self-interested actions of individuals.
2. To make our platform look global and successful from the outset, we made sure new users completed the country field in their profiles. As a result, their names, ages and country flags were automatically relayed to the front page of challengesЯus.com. It was designed as psychological reinforcement, pushing new members to sign up upon seeing that their countrymen were already members.

2. Meeting Claire

1. A kind of skin therapy using seawater, which is unusually popular in France.
2. When you have an internship at a high-caliber organization, the United Nations, your schedule is not exactly flexible.
3. A pejorative term Thai natives give to foreigners.
4. An aromatic, stir-fried rice dish with lime, fish sauce, cilantro and jasmine rice.
5. A nickname for Thailand that stuck thanks to a savvy marketing campaign by the Tourism Authority of Thailand (TAT) in the 1980s to attract tourists.
6. A term coined by the business press for the high-growth economies of Hong Kong, Singapore, South Korea and Taiwan, particularly from the 1960s up to the 1997 Asian financial crisis.
7. A clunky computer reservation and global distribution system for flight-booking owned by the Amadeus IT Group, based out of Madrid, Spain.

8. At the time of writing, marketing experts estimate that most Americans are exposed to around four to 10,000 ads each day, which includes outdoor and online advertising. Because of a lack of regulations, I am certain that Bangkok in the early 2000s was already reaching those astronomical numbers from offline advertising *alone*.
9. That is, "the line" separating the broad tri-media advertising of TV, radio and print from everything else.

3. Cole+Moore, Magnus and I

1. Though we were very much in love and ready to start a family together, Claire and I didn't get married until many years later.
2. In the world of consulting, the "big three" tend to focus on serious number-crunching and high-level management strategy (for which they command exorbitant fees). Moving down the hierarchy, smaller consulting firms often handle more of the in-the-trenches work, namely the specific and cultural "how" managerial questions within the larger "what" of a company's broader strategy—otherwise known as strategy implementation.
3. Engineering a company's "culture" to align employee mindsets and processes at all levels to accomplish upper management's company-wide vision.
4. This process used to be more manual; today, it is an automated, well-oiled process that has become foundational to online marketing.
5. Literally translated, *fika* is a "coffee and cake break," but in practice, it is a nearly religious ritual of socializing with food, an essential part of each day and a hallmark of Swedish culture.
6. A price-to-earnings ratio measures a company's stock's price per share divided by the company's earnings per share, used by investors to determine whether a company is overvalued or undervalued.
7. Something like Snapchat's geofilters of 2014 or Pokémon GO in 2016.
8. Like Apple's Notes app, released in 2012.
9. Like Rover in 2011 or Wag in 2015.
10. Fishbrain, a Swedish startup, had not launched yet. At the time of writing, they are the world's number-one fishing app.
11. Doctors Without Borders, the French non-profit offering international humanitarian aid in conflict zones and countries with endemic

diseases.

4. The Sun above the Clouds

1. An Initial Public Offering of stock is when a company's shares are available for the public to buy for the first time.
2. This was a flanking maneuver well known to even the most junior of consultants: adding up the years of experience a team had as individuals in total.
3. Short for Minimum Viable Product, an entrepreneurial concept coined by Frank Robinson and popularized by Steve Blank and Eric Ries in their book *The Lean Startup*, an MVP is a version of a product that is still useful to customers with the least amount of functionality included (with the goal of being first to market and using customer feedback for further product development).
4. Insight on just how unprepared Magnus and I were to start and run our own tech company.
5. The "orphan" he was giving us was a scrapped mobile game, the core of which included a code base for quests that users could complete; the idea was that we could repurpose the "quest" code for our challenges to save us some development time.
6. Much like you would see on the TV show *Shark Tank*.

5. Uninformed Optimism

1. A term for the exponential growth curve successful startups look for, wherein the early tinkering makes a line on a chart look like the head of a hockey stick up to a growth inflection point, where the line surges upward and looks like a long handle.
2. The length of time needed to turn a raw idea into a product widely available for sale.
3. For VCs, this means funding a lot of startups that are destined to flop for every one that hits; for startups, this criticism meant that we were developing too many different features and strategies in the search for a reliable revenue stream (our "prince").
4. An "innovative step" or "inventive step" is a requirement in most patent laws that a proposed product or invention is "non-obvious"

enough to merit intellectual property protection—and for any business, the more patentable it is, the more profitable.

5. Meaning that in exchange for an interest-free grant from a Swedish innovation fund, the company gave the government funder a block of stock options, which they could later cash in for a partial ownership stake in the company or sell to another interested buyer.
6. As of writing, approximately $23,000.
7. Sweden covers a wide expanse of land; it is also very tech-forward. To give an idea, Umeå is a seven-hour drive north of Stockholm; driving another seven hours north (still in Sweden) takes you to Kiruna, which is home to the European Space Tracking Network and Esrange, the European Space and Sounding Rocket Range. For its space-related infrastructure, Richard Branson's company Virgin Galactic has chosen Kiruna to host their own European spaceport.
8. I later learned that as a language, Swedish becomes simpler the further north you go. Having visited Kiruna once myself, every "yes" I heard there sounded like the sound made when sucking up a drink through a straw. This sound is so common in the northern hemisphere that historians theorize that it may be a vestige of Viking culture or a way of conserving energy in extremely cold weather.
9. Hypertext Preprocessor, a scripting language for web development. Strangely, PHP is a recursive acronym that stands for "PHP: Hypertext Preprocessor," meaning it includes itself.
10. Shortened name for Asynchronous JavaScript and XML, a set of technologies and languages for web development that include HTML and CSS.

6. A Lifeline from Caroline

1. The classic paradox in game theory that shows why two completely rational individuals might not cooperate, even if it appears to be in their best interest to do so.
2. In philosophy, a Weltanschauung is an intellectual construct that provides both a unified method of analysis for and a set of solutions to the problems of existence.
3. At the time, approximately $92,400.

7. Growing Pains

1. Though Sting still had a hand in it—in essence, Sting had deals and arrangements with various local office buildings that weren't direct Sting affiliates but had cut deals with Sting about rent. The offices would charge Sting's incubees less for their leases, and in exchange, Sting would send them potential lessees to keep all their units full.
2. Including *The Daily Drucker* from Peter F. Drucker, *Leading the Revolution* by Gary P. Hamel, *The Fifth Discipline* by Peter M. Senge, *The Empty Raincoat* by Charles Handy, *The Purple Cow* by Seth W. Godin and *Seizing the White Space* by Mark W. Johnson.
3. The Swedish equivalent of Albertsons or Kroger.
4. Sweden's national agency in charge of marketing the country for tourists abroad.
5. Maslow's hierarchy is a pyramid with more primal needs like food and water at the base and more abstract needs like self-actualization at the very top.
6. The World's Strongest Man of 1998.
7. A former professional tennis player who got to 32 on the ATP.
8. A social media campaign posted on social media between July to August 2014, challenging people to dump a bucket of ice water on themselves to promote awareness of Amyotrophic Lateral Sclerosis (ALS, a motor neuron disease).

8. Pivot

1. A situation where you are losing existing users faster than you are acquiring new ones.
2. Croesus was a sixth-century BC king of Lydia, an ancient kingdom in what is now modern-day Turkey.
3. Even in the notably gender-equal Sweden, men still outnumbered women in the startup world.
4. Python is a high-level, general-purpose programming language; among coders, both PHP and Python have their loyalists (though Python is growing in popularity while PHP is in relative decline).
5. Acronym for Linux, Apache, MySQL, PHP/Python.

9. Any Press

1. Needless to say, creating some synergy or an economy of scale by operating in two different lanes didn't really apply at our size—if anything, it was only a disadvantage.
2. Of course, in hindsight, they *did* get it—they just couldn't stomach the risk. How would we be able to execute the grand ideas we had? And what kind of money would that take?
3. It was not breaking news to me that the odds were against us. In fact, it was something Magnus had gone on about at length—that we had a better chance of making money by going to the casino, the latter being "about the same thing" in terms of risk-reward as starting a company. He had read all about it.
4. When I first met him in 2007, he had run me through all of its glorious details—Anders had always had a taste for the finer things in life.
5. The quote "It is not the strongest of the species that survives, but rather, that which is most adaptable to change" more accurately describes Charles Darwin's arguments in *On the Origin of Species by Means of Natural Selection* (1859), but Darwin never actually said it —it is a quote from Darwin paraphrased by Professor Leon C. Megginson in 1963. The first real use of "survival of the fittest" was by Herbert Spencer in *The Principles of Biology* in 1864. There, too, "fittest" meant "most adaptable to the environment."

10. Shooting from the Hip

1. Another important point was that if we had appeared even slightly similar to Cole+Moore from the start, we would have been doomed. Caroline would never have sponsored us, which would have translated into a much tougher start—and perhaps even no start at all.
2. Of course, sometimes we were forced to sign.
3. In the context of pay-per-click advertising, using a bot to pump up clicks is known as "click fraud" and is illegal—fortunately, what we did was not, as we were just manipulating the website's native algorithm for placement and not for money.
4. At least it would be heavily implied that we had worked together for anyone listening in.

5. Though audits are useful for some industries of scale, the length and detail of any audit document are simply horrendous and often a huge waste of valuable time. Their global cost to businesses each year must be phenomenal.
6. For the trickier questions, Google turned out to be very instructive.
7. A diagram of all the arranged links and nodes of our server system and how they interacted with one another.
8. Clean desk policy (or CDP) is a staple of corporate management strategy and is exactly what it sounds like: employees' desks must be kept clean and clear of all clutter (largely so employees wouldn't leave papers with corporate secrets lying around).
9. Internal Standards of Organization for quality management.
10. The Secure Sockets Layer (SSL) protocol, developed by Netscape in the mid-to-late '90s, is a cryptographic framework for safe communication between computers over a network. It was used in computer "handshakes," wherein two computers exchange and decode encrypted information ("keys") to establish trust and security. A '90s modem's signature chirps and blares are the coded and decoded sounds of two computers handshaking. SSL has since been replaced with the improved Transport Layer Security (TLS) protocol, though TLS's purpose is primarily still the same.
11. Single Sign-on, an authentication process that allows users to log in to multiple related but independent software systems. It is not at all related to SSL.
12. In retrospect, it is supremely ironic that we pretended to have an unstable connection in response to a question about our network security, though, of course, we didn't appreciate that at the time.
13. We thought of them a bit like supermarkets did diapers and other baby products—they sold at a loss, but they attracted families who made up those losses by filling their trolleys with other high-margin products.
14. A technique widely used in retail expressed as "odd prices." A classic example is something like "Supersize your meal for only €2.99!" In the above, consumers tend to perceive €2.99 as spending only two euros when they're really spending one cent less than three.
15. Adding all costs of offering a product or service and adding a percentage on top of it as profit.
16. Pricing a product based solely on its perceived value to the customer, for which there is no easy metric or benchmark.

17. The official definition of price elasticity is the relationship between a change in the quantity demanded of a product or service and a change in its price. In short, if demand is elastic, revenue is gained by reducing the price; if demand is inelastic, the opposite is true. For example, if Apple were to raise the price of iPhones, many would still buy them—people need smartphones, and Apple's product is widely viewed as the best. If the price of Campbell's Soup were to rise, however, there are so many good alternatives that people would likely switch to cheaper options in a heartbeat.
18. Or top euro, technically.

11. Business Unfiltered

1. Our strict adherence to the "time is money" idea was also inherited from our consulting days. Of course, there was a flipside to our fanatical devotion to efficiency: Magnus and I often debated the value of pursuing or not pursuing an opportunity for hours.
2. Though I didn't recognize her, Magnus told me that she was Swedish singer, blogger and presenter Annika Sandström, who had presented the previous year's Melodifestivalen, an annual song competition that determined Sweden's representative for the Eurovision Song Contest.
3. The fourth industrial revolution, wherein the Internet of Things (IoT), AI and machine learning will be integrated into production facilities to create "smart factories."
4. Software as a service.
5. Armenia has a rich and fascinating culture dating all the way back to 4000 BC. Bordered by Turkey, Georgia, Azerbaijan and Iran, the country faced countless hardships throughout its history, including a widely unrecognized genocide as part of the Ottoman Empire in 1915, a massive earthquake in 1988 that killed more than 25,000 people and an unprecedented economic collapse after breaking away from the Soviet Union in 1991.
6. Epinephrine is a neurotransmitter released by the locus coeruleus in the brainstem. Corticotropin is a hormone made and secreted by the pituitary glands, and its main purpose is to signal an increase in production of cortisol, the stress hormone.
7. I'm not sure he recognized me, but I recognized him—Magnus and I had met him at the Spotify party. He was one of the people who had

asked to be on our newsletter list.

8. An Italian sportswear brand founded in Turin.
9. Megarian banded centipedes, the most common of its order found in the Mediterranean region, and yes, I know people who do this.
10. And I quite enjoyed it, to be perfectly honest.
11. Both from talks I had given as a consultant and from all the wonderful TED Talks I was addicted to (one of my *otakus*).
12. Ballmer had commanded an incredible level of energy by yelling "Developers, developers, developers!" in a room jam-packed with thousands of Gates-like, introverted techies—not the easiest crowd to arouse.
13. A zone of 26 European countries that have functionally abolished their borders for other members, so named for being signed in Schengen, Luxembourg in 1985. The zone applies to nearly all Western Europe as well as Greece but currently excludes Bulgaria, Croatia, Cyprus, Gibraltar, Ireland and Romania (and, of course, the UK).
14. The latter was my way of explaining what entrepreneurship was like.
15. A concept outlined in a book by the same name, the innovator's dilemma was the conceptual problem that at first, disruptive technologies did not satisfy the needs of any part of the market (not even the high-end) because nobody knew about them yet. Thus, large companies tend to ignore innovative new technologies for too long—because on a long enough timeline, innovative technology surpasses old tech, finds a market and becomes inexpensive enough to deploy at scale. Unfortunately for large corporations, it is often too late by that point to invest and adapt fast enough to catch the early-adopting companies who were there first.
16. A way of valuing an asset or security based on the expected future cash flows it will produce.
17. Since I hadn't known the term in my first failed pitch meeting at Sting, I had to include it.
18. Boolean logic is a form of algebra centered on the idea that all values are either true or false. It has been around since the 1800s, after being initially proposed by George Boole, a renowned mathematician. From a programming perspective, Boolean logic is the concept of breaking complex decisions into a series of true and false questions.
19. I was reminded of Henry Ford's solution to the problem: announcing that his customers could have their cars painted any color they wanted as long as that color was black.

12. Malediction

1. Caroline, of course, was analyzing our company from a consultant's point of view; Magnus, on the other hand, was looking at it like a founder.
2. I recognized the line from the DreamWorks movie *Madagascar*, which our children had been watching repeatedly in preparation for the voyage while Claire and I attended to our evening chores.
3. Mozambique was colonized by the Portuguese Empire in 1498, beginning with expeditions led by Vasco de Gama.
4. A Swedish-Swiss corporation mainly involved in manufacturing heavy electrical equipment and developing robotics and automation technologies.
5. To potentially include things such as securing rights to exclusive distribution, white labeling our platform and actively participating in our company through minority shareholding.
6. When he was heading IKEA, the biggest multinational furniture retail company in the world, the legend goes that Kamprad never flew business class despite a personal net worth in the tens of billions of dollars.
7. To carry out his plan, the idea was to have Bishop Brennan come to the island's parochial house to witness his own image, which had, as a "miracle," appeared on the tiny skirting board of Ted's room. When the rather old Brennan leaned forward to try to observe the smallish image, Father Ted would kick him. The grand idea was because no one in the real world would dare kick the bishop in the behind; not even the bishop would believe what had just happened. He would have thought he had just dreamt it all. Father Ted would then continue as if nothing had happened. The plan and the execution were perfect, but the outcome was not the one Father Ted had hoped for.
8. Eventually, we told the story to every new employee as part of their onboarding program. It made us vulnerable and fed into our narrative of being "cowboys in suits" who shot from the hip, failed fast and forward and went all-in.
9. In *The Protestant Ethic and the Spirit of Capitalism*, Weber, a German sociologist and political economist, attributed the rise of market-driven capitalism to the values of "hard work."

10. Carpe diem translates to "seize the day," but Horace didn't mean it in the "rise and grind" sense that entrepreneurs today might. Horace came from the school of epicureanism, an essentially hedonistic ideology of living only for the moment and not thinking about the future. I doubt Horace and Weber would have gotten along well.
11. The godfather theorist of market capitalism and the man who coined the "invisible hand" of markets.
12. For an idea of the kind of content I was absorbing, search "Will Smith Motivational Video" on YouTube.
13. Recent scientific research seems to conclude that when humans engaged in so-called persistence hunting, they could outrun antelopes and other game. Not because of their speed, of course, but because of their stamina and their ability to cool themselves efficiently like any other mammals on earth.
14. In fact, there is even a correlation psychologically between one's level in the organizational chart and one's level of risk aversion.
15. As the old saying goes, "No one ever got fired for choosing IBM."
16. We gradually put it together that the first of these cowboys was Dennis, the HR manager at our massive healthcare company. The fact was if Dennis's main goal was to keep his job and advance in a company, then he'd been a fool to buy our services. It raised other questions: Did he really believe in our vision? Or had he only hired us as an act of corporate seppuku? We wanted answers, but he had left his company and fallen off the map. We hadn't been in contact for months.

13. All Hell Breaks Loose

1. Kaapsehoop or Kaapschehoop is a town in the Mpumalanga province (formerly Eastern Transvaal), South Africa. It is situated at 1486 m (4878 ft.) above sea level on the Highveld escarpment, about 25 km from Nelspruit, the capital of Mpumalanga. Nelspruit is about 200km away from Maputo, Mozambique.
2. 104 F.
3. Malediction theories aside, one-on-one interviews with senior leaders had always been an integral part of our processes. "Customer churn" was a serious topic for any organization, but it was particularly important for startups because it was much easier (and less costly) to keep

existing customers than it was to gain new ones. Clearly, as a young company, you want to stay close to your buyer to further your understanding of why they chose you in the first place.

4. The first time I ever actually "met" Dennis was at Cole+Moore, when I had approached him trying to sell the firm's change management services, which I never managed. I wasn't sure he even remembered that interaction, though I'd remembered it when Dennis first came on as a client at ChallengesЯus.
5. Military term for Joint Direct Attack Munition, a bolt-on guidance system that can turn "dumb" bombs into low-cost smart bombs. In essence, Magnus was telling me that we had a missile coming directly at us.
6. Radio code for "what is your location?"
7. Fortunately, we did have such a clause in our contract allowing us to do so.
8. Bastille Day, the anniversary of the Storming of the Bastille on July 14th, 1789, perhaps the biggest event in the French Revolution, has been celebrated annually ever since with a full-fledged military parade.
9. This infuriated us at the time, as money that could've helped us tremendously had instead essentially been set on fire over a scam. It left a sour taste in my mouth about cryptocurrency and blockchain technology in general, though the intervening years have shown that it is a phenomenon that is likely here to stay.
10. The former Chief Business Officer of Google X, a unit of Alphabet that works on self-driving cars and sends high-altitude, super-pressured balloons into the stratosphere to beam the internet over rural areas.

14. A Platform for Growth

1. And there had been many, many minor iterations between.
2. Of course, what Rumsfeld considered "known knowns" and "unknown unknowns" as those concepts mapped onto the invasion of Iraq and Iraq's alleged possession of WMDs is a bit nebulous, but in any other context, it remains an interesting point about strategic decision making.

3. Including corporate communiqués, press releases, training materials and more.
4. How hard could it be, right?
5. I was not as geared up as some runners, but I did buy myself a top-of-the-line Garmin running watch. Beyond helping performance, it was a signal that announced to strangers *I'm serious about running.* In the years since, wearing it has opened business opportunities; many executives I meet are high achievers at work but also on the road, in the pool and, sometimes, in all three disciplines.
6. We had an impressive brag sheet that included several of the most admired companies in the world. Sure, we had never actually delivered a comms platform solution to them, but what did that matter?
7. Lewin's model of psychological and organizational change came from the 1940s, so I supposed that the men in the audience would've been all too familiar (and comfortable) with it.
8. Which I did on the off chance that I could somehow influence the course of events through my thoughts alone.
9. It may sound like a lot, but it really wasn't—it was just enough to allow us to stay in the game.

15. The Party Is Over

1. There was also a looming fourth possibility we didn't dare speak of: that of crashing and burning miserably if we couldn't quickly fill our coffers.
2. A hopefully accurate estimate of a company's valuation before receiving funding vs. after receiving it.
3. The Growth Share Matrix, popularized by the consulting firm BCG, sorts companies by market share (high or low) and growth potential (high or low). Low market share and low growth is a "dog" company, unworthy of investment. Its opposite is a high growth, high market share company, a "star" you could throw money at. Though we were perhaps not yet a star, we were likely now a "question mark" (high growth—high*ish* in our case—low market share) with potential to become a "cash cow" (low growth, high market share).
4. Business to consumer, our first iteration.
5. In 1991, Geoffrey Moore wrote *Crossing the Chasm: Marketing and Selling High-Tech Products to Mainstream Customers*. A chasm exists

because it takes a lot of effort post-selling a product to early adopters to take it to the masses.

6. Oh, to be so naïve again.
7. The "stickiness" of a product or service refers to how much a customer wants (or has) to keep coming back to use.
8. A term often used in startup circles in reference to Michael Jackson's "Pretty Young Thing," the smash hit from his 1982 album *Thriller*.
9. A fundraising round wherein a company fails to match or preferably beat the amount of money they raised in the round before, which is widely considered startup suicide.
10. As the saying goes, "The house always wins."

16. The Last Pivot

1. "This too shall pass" is a popular saying first recorded in medieval Persian Sufi poems, though it has since become widely popular around the world as a reflection on the ephemerality of the human condition.
2. An intended pun which I'm not sure he picked up.

ACKNOWLEDGMENTS

This book would never have become a reality without the encouraging words from the people, friends and complete strangers to whom I pitched the story. They are the ones who told me I should turn this crazy entrepreneurial journey into a book. Over time, the idea grew on me.

Because there would have been no book without ChallengesЯus, I would like to take the opportunity to thank the various people whose lives directly touched mine as they got involved in my startup journey, some of whom were introduced as characters in this book.

First, a big thank you to David and Frida for being the best at what you do. Clearly, without you there, there would have been no company at all.

A very special thanks to Peo, Pär and the whole team at Sting for believing in us so early on.

My gratitude extends to Donnie and Debbie for the roles they took in the early days of the company.

A huge thank you to Katrin, Fredrik and the entire crew at C&M, without whom ChallengesЯus might have stayed a mere PowerPoint presentation.

I am thankful to the thousands of early adopters whom I

don't know personally but who made us believe in the project when we had absolutely no idea what we were doing.

Thanks to the many celebs who accepted the challenge and put their names on the line.

A special thanks to Eric Larsen, Decorey Jones, Sara Nyberg, David Persson and my sister Aude for taking the lead.

We would, of course, never have become a real company without our initial paying corporate customers and partners. I am especially appreciative to the very first ones—those with whom I have come to build a unique relationship. Among them are Chantal, Catherine, Björn, Elin, Isabelle, Anders, Guillaume, Katri and Jo.

I would also like to thank Kaisa, Marju, Ed, Mark, Laura, Charlotte, the other Jo, Michaela, Karunesh, Linus, Andrea, Ulrich, and Tansu for having played their part in our venture.

Thank you to Matthew at the University of Leicester for making me feel like a professor for a day.

I am particularly indebted to my wife, Claire, a superwoman without whom my entrepreneurial journey—including this book—would never have even started. It must have been tough to live with me year after year, through the ups and downs first of ChallengesЯus and then of xtrnlDesk—and yet, in the end, there is still no house by Lake Como (not yet at least). So, thanks, Smooch, for believing in me all along and for taking care of our kids the million times I had to lock myself in the study to finish "something important."

Thanks to my kids for keeping up with a stressed Dad who was not always 100 percent present.

I would like to thank those who have given me their precious time to read early drafts of this book and who have given me their invaluable feedback. Special thanks to Maria de Zaragoza, Nivin, Erik P., Knut, Julia and Matthew. I would like to give a shout-out to Brian for relentlessly going over my last draft to find ways to improve the final product. Thanks mate.

I have been very lucky to meet, literally out of the blue, (probably the universe at play) with Rajesh. Rajesh believed in the story so much that he introduced me to Anna and her publishing company, Launch Pad Publishing. I am very lucky to have had Ryan as my editor. He has been absolutely amazing—a brilliant intuitive sense of what needed to be said (and what did not). His relentless search for perfection has considerably improved the book. Thank you, Ryan, I've very much enjoyed working with you.

Finally, I would like to thank—from the bottom of my heart—my then-partner Magnus. In the early days of ChallengesЯus, Magnus and I sat over a weekend to iron out the details of the very first official document we ever produced for the company: our shareholder agreement. Beyond the fun topics of tag-along and drag-along rights governing what happens when a company gets purchased or shares get sold, there was a clause that considered what would happen to the company if any one of us was to be physically or mentally incapacitated or, God forbid, pass away. It felt surreal. We were literally scenario planning and making provisions for the worst possible thing that could ever happen to anyone. What were the odds, really?

Little did we know that four years down the road, Magnus would live through any parent's worst nightmare: the loss of his beloved daughter, Lovisa, at the age of three. Lovisa was a wonderful little girl who was part of our adventure. Both Magnus and I were young parents, and I will forever remember those numerous Skype video calls where she sat on her Daddy's knees while we strategized. She left this world very unexpectedly from an extremely rare genetic disease. It was so abrupt and so horribly painful. Magnus and his wife Josefine were devastated; who wouldn't be? To lose a child is the worst possible thing that can happen to a parent. Magnus had to be away from the business for quite some time to deal with his sorrow, and I cannot fathom what he must have gone through. Just being a firsthand witness was traumatic for me (and still is, in so many ways).

Yet, after 10 months or so, he made a case for coming back to the company. I had doubts. I wondered how he would possibly be able to take the stress we were constantly enduring in addition to his own personal pain. I would not have been able to, that's for sure—but he did, and he did it well. His days were filled with suffering, but he worked hard on himself and brought all he had left to ChallengesЯus.

We had started our adventure together, and he wanted to go all the way to the finish line together. That was our initial deal. It is remarkable what commitment, passion and friendship can lead one to do. I never told him, but from that day he came back, he had become a hero to me. Champions don't quit. He never did. This journey would never have been the same without you buddy. Thank you! You are a real trooper. Roger. Out.

ABOUT THE AUTHOR

Arnaud Henneville-Wedholm is a consultant, lecturer and the head of sales and business development at GLOBHE. He is also the founder of multiple startups, including internalDesk, a SaaS platform for enterprise collaboration, where he served as COO. Henneville-Wedholm's interest areas include entrepreneurship, neuroscience, resilience and making the world a better place. He lives in Stockholm, Sweden with his wife and three children. *How Hard Can It Be* is his first book.

For startups:
For startup incubators looking to help their entrepreneurs (with the book, pep talks or coaching interventions) or startup founders and employees who want to accelerate their success, or:

For corporate events:
For keynotes or sessions on topics of innovation, mindset, culture, intrapreneurship, employee engagement and change, please reach out to hello@howhardcanitbethebook.com.

For bulk purchases or custom books (with intros by your CEO, for example), please email

hello@howhardcanitbethebook.com.

* * *

It is everyone's duty to illuminate others' paths when we can. If this book has inspired you, please share it with someone you know will benefit from it.

Connect with Arnaud Henneville-Wedholm:

facebook.com/How-Hard-Can-It-Be-108402478235845
instagram.com/arnaudhenneville
linkedin.com/in/henneville-wedholmarnaud

Made in the USA
Monee, IL
13 September 2026

63177765R00156